Automotive
Engine
Metrology

Automotive Engine Metrology

Salah H. R. Ali

PAN STANFORD PUBLISHING

Published by

Pan Stanford Publishing Pte. Ltd.
Penthouse Level, Suntec Tower 3
8 Temasek Boulevard
Singapore 038988

Email: editorial@panstanford.com
Web: www.panstanford.com

British Library Cataloguing-in-Publication Data
A catalogue record for this book is available from the British Library.

Automotive Engine Metrology
Copyright © 2017 Pan Stanford Publishing Pte. Ltd.

ISBN 978-981-4669-52-8 (Hardcover)
ISBN 978-1-315-36484-1 (eBook)

Printed in Canada

To Egypt and the World

To the soul of my parents, my mother, Hakemah Ahmad El-Banna, and my father, Hamed Ramadan Ali

To my respected professors, and teachers

To my lovely wife, Hayam, our son, Amr, and our daughter, Maryam

To all those who helped me in my career and to my colleagues at the National Institute for Standards, the Academy of Scientific Research and Technology, and the Ministry of Higher Education and Scientific Research in Egypt

To my postgraduate and undergraduate students, production and quality engineers, automotive engineers, metrology engineers, each researcher and science student, and all interested

Salah H. R. Ali

Contents

Preface   xiii

PART 1: INTRODUCTION

1. **Introduction**   **3**

　　1.1　Automotive Engine Metrology   4

　　1.2　Engineering Metrology   10

　　1.3　Quality Challenges in Automotive Engineering   14

　　1.4　Metrology Laboratory   16

　　1.5　Automotive Engineering   17

　　　　1.5.1　Types of Automotive Engines   18

　　　　1.5.2　Performance of Automotive Engine   21

　　1.6　Conclusion   23

PART 2: ADVANCED METROLOGY TECHNIQUES

2. **Advanced Measurement Techniques in Surface
   Metrology**   **27**

　　2.1　Advanced Measuring Techniques   28

　　　　2.1.1　Mechanical Contact Stylus Techniques   28

　　　　　　2.1.1.1　CMM coordinate technique   29

　　　　　　2.1.1.2　Roundness instrument   33

　　　　　　2.1.1.3　Roughness measurement technique   38

　　　　2.1.2　Optical Measurement Techniques   41

　　　　　　2.1.2.1　White-light interference microscopy   42

　　　　　　2.1.2.2　Confocal optical microscopy   44

　　　　　　2.1.2.3　Confocal white light microscopy   45

　　　　　　2.1.2.4　Scanning electron microscopy   46

　　　　　　2.1.2.5　Digital holography technique   48

　　2.2　Non-Optical Measurement Techniques   49

　　　　2.2.1　AFM Technique   50

2.2.2   3D-CT Technique 52

2.3   Overlapping, Limitations, Sampling, and Filtering
of Existing Techniques 54

2.3.1   Overlapping 54

2.3.2   Limitations 55

2.3.3   Sampling and Filtering 56

2.4   Surface Characterization 57

2.4.1   Applications in the Mechanical Engineering 57

2.4.2   Other Applications 61

2.5   Uncertainty 66

2.6   Conclusion 68

PART 3: PERFORMANCE OF CMM METROLOGY TECHNIQUE

3. **Characterization of Touch Probing System in CMM
Machine** **83**

3.1   Types of CMM Probes 84

3.1.1   Hard Probe 84

3.1.2   Trigger Probe 84

3.2   Analytical Model 87

3.2.1   CMM Probe Ball Tip Error 87

3.2.2   Results of Analytical Model 89

3.3   Experimental Work 90

3.3.1   Verification of CMM Stylus System 91

3.3.2   Experimental Procedure 91

3.3.3   Parametric Study of Stylus Design 93

3.3.4   Measurement Density 93

3.3.4.1   Stylus tip size 4.0 mm 95

3.3.4.2   Stylus tip size 2.5 mm 95

3.3.4.3   Stylus tip size 1.5 mm 95

3.4   Analysis of the Obtained Uncertainty 96

3.5   Experimental Results and Discussions 97

3.5.1   Effect of Probe Stylus Tip Size 98

3.5.2   Effect of Probing Speed 102

3.6   Conclusion 102

4.   Error Separation of Touch Stylus System and
CMM Machine   **107**

   4.1   Experimental Work   109
         4.1.1   Verification of CMM Machine   109
         4.1.2   Parametric Study of CMM and Stylus Design   109
   4.2   Analysis of Experimental Results   111
   4.3   Validation of Experiments   117
         4.3.1   Total Measurement Errors   117
         4.3.2   Stylus System Errors   118
         4.3.3   CMM Machine Errors   119
         4.3.4   Other Measurement Errors   120
   4.4   Conclusion   120

5.   Measurement Strategies of CMM Accuracy   **125**

   5.1   Introduction   125
   5.2   Background and Motivation   128
         5.2.1   Types of Errors   129
         5.2.2   Fitting Algorithm   130
   5.3   Experimental Work   130
         5.3.1   General   130
         5.3.2   Dynamic Calibration of Stylus System   131
         5.3.3   Test Procedure   132
   5.4   Result Presentation and Discussion   133
         5.4.1   Probe Scanning Speed 5 mm/s   133
         5.4.2   Probe Scanning Speed 10 mm/s   135
         5.4.3   Probe Scanning Speed 15 mm/s   136
         5.4.4   Probe Scanning Speed 20 mm/s   137
         5.4.5   Probe Scanning Speed 25 mm/s   139
         5.4.6   Probe Scanning 30 mm/s   140
         5.4.7   Probe Scanning 35 mm/s   141
         5.4.8   Probe Scanning Speed 40 mm/s   142
         5.4.9   Probe Scanning Speed 45 mm/s   143
   5.5   Statistical Analysis   145
         5.5.1   Standard Deviation Average of Roundness
                 Measurement Error   145

5.5.2   Roundness Error of Scanning Speed
Response   146

5.6   Conclusions   148

**6.   Validation Method for CMM Measurement
Quality Using Flick Standard   153**

6.1   Introduction   153

6.2   Experimental Work   155

6.2.1   Dynamic Verification of Probing System   155

6.2.2   Flick Standard Artifact   156

6.2.3   CMM Measurement Procedures   157

6.3   Measurement Results and Discussion   158

6.3.1   Least Square Fitting Technique   158

6.3.2   Minimum Element Fitting Technique   161

6.3.3   Minimum Circumscribed Fitting Technique   164

6.3.4   Maximum Inscribed Fitting Technique   167

6.4   Statistical Analysis   169

6.4.1   The Error in Diameter Measurement   170

6.4.2   The Error in Roundness Measurement   172

6.4.3   Uncertainty Evaluation   175

6.4.3.1   Repeatability   175

6.4.3.2   Resolution   175

6.4.3.3   Indication error   175

6.4.3.4   Temperature   175

6.5   Conclusions   176

PART 4: PERFORMANCE OF TALYROND METROLOGY TECHNIQUE

**7.   Factors Affecting the Performance of Talyrond
Measurement Accuracy   183**

7.1   Introduction   183

7.2   Background and Motivation   185

7.2.1   Fitting Filters   186

7.2.2   Fitting Spectral Wave Responses   186

7.2.3   Fitting Algorithms   187

7.2.4   Types of Errors   188
7.3   Experimental Work   189
7.4   Results and Discussion   191
      7.4.1   The Effect of Fitting Filters   191
      7.4.2   The Effect of Gaussian Filter and Fitting
              Techniques   193
      7.4.3   The Effect of 2CR Filter and Fitting
              Techniques   195
7.5   Analysis and Estimation of Roundness Accuracy   197
7.6   Conclusion   199

PART 5: METROLOGY IN AUTOMOTIVE ENGINES

**8.   Metrology as an Inspection Tool in New or Overhauled
       Water-Cooled Diesel Engines   205**
8.1   Introduction   205
8.2   Engine Inspection Program   207
      8.2.1   Engine General Specifications   208
      8.2.2   Cylinder Liner Inspection   208
      8.2.3   Valve Lapped Area Inspection   210
      8.2.4   CMM Verification   210
      8.2.5   CMM Measurement Strategy   211
8.3   Experimental   211
      8.3.1   Cylinder Block Measurements   212
      8.3.2   Cylinder Head Measurements   212
      8.3.3   Valve Measurements   214
      8.3.4   Piston and Ring Measurements   215
      8.3.5   Measurement of Engine Performance   216
8.4   Uncertainty in Measurements   216
8.5   Results and Discussion of Engine Inspection   219
      8.5.1   Results of Dimensional Deviations   220
      8.5.2   Results of Form Deviations   220
      8.5.3   Results of Location Deviations   222
      8.5.4   Results of Engine Compression Pressure   222
8.6   Conclusion   225

9. Metrology as an Identification Tool for Worn-Out Air-Cooled Diesel Engine   **229**

9.1   Introduction   229

9.2   Cylinder Forces and Surface Measurements   232

    9.2.1   Dynamic Friction Force   232

    9.2.2   Surface Geometry Measurements   233

9.3   Uncertainty Assessment of Measurements   235

9.4   Results and Discussion   237

    9.4.1   Out-of-Roundness Measurement Results   237

    9.4.2   Concentricity Measurements   239

    9.4.3   Out-of-Straightness Measurement Results   239

9.5   Conclusions   240

10. Surface Metrology in Engine Quality   **245**

10.1   Engine Quality Using Metrology Techniques   247

    10.1.1   CMM Metrology Technique   247

    10.1.2   AFM Metrology Technique   249

    10.1.3   Scanning Electron Microscopy Technique   250

    10.1.4   Transmission Electron Microscopy Technique   251

10.2   Tribological Behavior   252

10.3   Coated Surface Characterization   253

10.4   New Applied Technology in Engine Coating Surfaces   258

10.5   Machining Characteristics of Engine Cylinder Surface   260

10.6   Conclusion   262

PART 6: CONCLUSIONS AND FEEDBACK FOR FUTURE

11. Conclusions, Recommendations, and Future Work   **269**

11.1   Conclusions   270

11.2   Recommendations   271

11.3   Future Work   272

Index   273

Preface

Advanced soft metrology techniques play an important role in improving the *quality* and *function* of automotive engines with regard to both manufacturing and diagnostic processes. Advanced accurate and precise measurement techniques are based on two fundamental approaches: hard measurement techniques and soft measurement techniques. Advanced soft computing measurement techniques include a coordinate measuring machine (CMM), Talyrond roundness tester, surface roughness device, interferometric methods, confocal optical microscopy, scanning probe microscopy, and computed tomography technique at the micro- and nanometer scales. Now, utilizing the CMM or the Talyrond machine is a challenge for advanced coordinate metrology in modern engineering applications, especially in automotive and aerospace industries. Deviation from dimensional tolerance or geometrical features can produce a number of engineering problems, vibration, frictional wear, noise, material fatigue, and failure. The basic function of the CMM is to measure the actual dimension and geometrical shape of an object according to the ISO and evaluate the collected data using the metrological aspects of size, form, location, and orientation.

In this book, we focus on advanced coordinate measurement machines and their performance with respect to accurate and precise measurements for automotive engine metrology. The book is organized into six parts. The first part presents the general introduction, the objective of the book, and its usefulness for academic scientists and professional and general readers. The second part introduces the important industrial subject of advanced soft measurement techniques for dimensional and surface metrology in micro- and nanometer scales. The third part discusses the performance and error analysis methods of the CMM as a new common technique for dimensional and surface metrology in the industry. The fourth part studies error analysis and roundness determination using the Talyrond technique. The fifth part discusses the inspection and diagnosis of new, overhauled, and worn-out automotive engines using the CMM technique. It also discusses the applications of surface metrology in quality control for automotive engines. New technologies for engine coating and surface characterization are also presented. The last part, Part 6, discusses the developments in the field and future prospects.

It is hoped that the book will encourage the development of techniques in instrumentation metrology for automotive engines and strengthen readers' understanding of the importance of metrology in automotive engines.

Salah H. R. Ali, PhD
Professor Doctor Engineer
Engineering and Surface Metrology Department
National Institute for Standards, Giza, Egypt
April 2017

PART 1
INTRODUCTION

Chapter 1

Introduction

Automobiles are considered as a lifeline in all matters of our daily lives. The automotive industry is the most sustainable, investment-friendly, and profitable field all around the world, especially in developed countries. About 80% of the world industries and their development are applied in the automotive industry. On the other hand, nowadays, one cannot live without the automotive transportation method. For that, scientists continuously explore necessary scientific methods in metrology to improve the quality of automotive performance. Therefore, metrology plays a very important role in the automotive industry, and at the same time, the progress in the automotive industry plays a very important role in the development of metrology techniques.

Metrology is a measurement science, and it finds applications everywhere in the daily life. In this regard, this author emphasizes, "any progress in the metrology is considered a major cause of actual progress in the automotive industrial technologies and vice versa."

The focus of this book is the concept of automotive engine metrology—in other words, how we can harness the advanced metrology techniques for developing automotive engine technologies? To achieve this goal by a simple scientific way, the author gives a brief introduction of attractive and interesting subjects through the important following topics: automotive engine metrology, which is covered through important research

Automotive Engine Metrology
Salah H. R. Ali
Copyright © 2017 Pan Stanford Publishing Pte. Ltd.
ISBN 978-981-4669-52-8 (Hardcover), 978-1-315-36484-1 (eBook)
www.panstanford.com

topics through the next 10 chapters in this work, automotive engineering, engineering metrology, quality challenges in automotive engineering, and metrology laboratory.

1.1 Automotive Engine Metrology

Advanced soft metrology techniques play an important role in improving the quality and performance of automotive engine in both manufacturing and diagnostics processes. Advanced accurate and precise measurement techniques involve one of the two basic approaches: the hard measuring techniques and the soft measuring techniques. The advanced soft computing measuring techniques include coordinate measuring machine, Talyrond roundness tester, surface roughness device, interferometric methods, confocal optical microscopy, scanning probe microscopy, and computed tomography technique in the different levels of measurement scale. Nowadays, the coordinate measuring machine (CMM) or the Talyrond roundness measurement tester is one challenge for advanced coordinate metrology in modern engineering applications, especially in automotive and aerospace industries. Out of dimensional tolerance or geometrical features can produce a number of engineering problems such as vibration, frictional wear, acoustic noise, material fatigue, and failure and accordingly affect the cost. The basic function of the CMM is to measure the actual dimension and geometrical shape of complex objects according to ISO standards and evaluate the collected data using metrological aspects of size, form, location, and orientation. The ability of the Talyrond tester as a height precise mechanical device lies in the importance of measuring the roundness for cylindrical and spherical parts with very high accuracy of up to nanometer.

In this book, we focus on the advanced coordinate measuring machines and the parameters that affect their performance in the accurate and precise measurements for automotive engines. Moreover, the verification method of the CMM probing system and the CMM measurement quality are discussed in detail to ensure the confidence in the automotive engines' efficiency. The book—organized in six parts comprising 11 chapters—sheds light on and discusses the important issues in automotive engine metrology.

The first part presents the general introduction, book contents and the aim of the book to display the inevitable necessary relationship between automotive engines and the metrology through the quality in the perimeter of interest for the academic scientists, professional engineers, engineering students, and general readers.

The second part introduces the necessary important industrial subject for advanced soft dimensional measuring techniques in micro- and nanometer scales (Chapter 2). Chapter 2 is dedicated to dimensional and surface metrology. The need for accurate dimensional measurements and quality engineered surfaces has become a necessary requirement to meet the challenges of modern technologies. Thus, advanced precise and accurate measurement techniques play a vital role in improving the function and quality of engineering products. The author discusses the advanced precise and accurate measurement techniques in terms of two basic approaches: the hard metrology techniques and the soft computing metrology techniques. The advanced soft metrology techniques include coordinate measuring machines such as roundness Talyrond machine, surface roughness devices, and optical microscopes. On the other hand, to complete the image, a new technical committee in ISO standards in the field of dimensional and geometrical product specifications and verification is established to address characterization issues posed by the areal surface texture and new measurement techniques. Here, different classification schemes of major advanced soft measurement techniques and their applications in industrial dimensional and surface metrology are reviewed. Moreover, current techniques, future trends under development and ISO strategies in this area are discussed.

Part 3, comprising four chapters, discusses the probing system, errors analyses, verification, and quality in the measurement accuracy of CMM performance, as commonly used in the newly processes for dimensional and surface metrology in the industry. Chapter 3 presents the characterization of probing system in CMM machines. The necessity of improving the quality of products and the ever-increasing demand for reducing the production duration resulted in the search for more accurate and faster implementation methods. Touch trigger probes are widely used on most commercial CMMs. This chapter studies the effect of the dynamic root errors on surface measurements using three different types of touch trigger probes attached to a bridge-type-CMM. Unforeseeable dynamic root errors of a ductile touch trigger probing system have been characterized theoretically and experimentally as well. The results employ in validating an analytical two-dimensional-model (2DM) of stylus tip to be developed to demonstrate the capability of such approaches of emphasizing the root error concept and to evaluate the accuracy of the CMM measurements. Experiments were conducted and the results were analyzed in order to investigate the effect of the dynamic root errors in the light of the probe scanning speed at different stylus tip radii. Variations in the mass and/or geometry of the stylus have their consequent effects on its inherent intrinsic dynamic characteristics that in turn would cause relevant systematic root errors in the resulted measurements. 3D bore cylindrical surface form undulations were measured by employing a probe with a trajectory of internal surface diameter for a standard reference test gauge ring. Regression analysis was applied on the results of measurements density distribution, and the uncertainty of measurement repeatability was then evaluated and graphically presented. The results were investigated and optimum strategic measurements parameters could thus be derived to ensure foreseeable accurate and precise results.

Chapter 4 presents very interesting analysis and discussion on the error separation of the stylus system and the CMM machine in detail. CMM is widely used for a large range of accurate and precise dimensional measurements. The micro-scale measurement is expected to show an ever-increasing performance in term of root error separation for stylus system and CMM machine accuracy. The experimental work in this chapter aims to study the effect of dynamic root errors at different undulations per revolution (UPR) response of artifact measurement using selected two types of CMM touch-triggering stylus. The direction of the stylus and stylus speed parameters were adapted and utilized throughout the course of experiment. The results were investigated using the Fourier analysis to ensure foreseeable accurate and precise results of CMM machine and stylus errors. Some specific error equations for the stylus system and machine responses have been postulated and analyzed empirically. The results also include validating the experimental investigation to predict accuracy using PRISMO-Bridge-CMM-type in NIS.

As CMM becomes one of the main requirements in precision engineering for advanced industries for design and troubleshooting and more for scientific facilities. The CMM data analysis software can contribute significantly to the total measurement errors and by extension on the product quality. The error characteristics in the CMM software are very important from the metrological point of view to find an optimum fitting solution. The final accuracy of a work piece is influenced by many different factors. Chapter 5 illustrates how academic scientists, metrology engineers, and future engineers stand in front of and can evaluate the CMM measurement accuracy using standard rings. In this chapter, the fitting software methods and styles touch probe scanning speed factors for three different transverse circles of roundness measurement errors are studied experimentally. The tests have been performed to examine the problem of how to generate reference data sets for cylinder circle measurements. Some error formulae have been postulated to correlate the roundness measurements within application range. These reference data sets are presented to help the CMM designer and operator to get the best fit for roundness measurements. Chapter 6 introduces a new verification method of CMM measurement accuracy using the Flick standard. The verification of CMM in dimensional metrology is very important task to ensure the quality of manufacturing processes. The new experimental verification method of CMM performance is designed, performed, and developed by using a reference standard artifact at the National Institute for Standards (NIS) in Egypt. The measurement errors of corresponding geometric evaluation algorithms and probe scanning speeds are obtained through repeated arrangement, comparison, and judgment. The experimental results show that the roundness error deviation can be evaluated effectively and exactly for CMM performance by using the Flick standard artifact. Some of influencing quantities for diameter and roundness form errors may dominate the results at all fitting algorithms under certain circumstances. It can be shown that the coordinate measurement at 2 mm/s is the best strategic parameter to satisfy the high level of accuracy in the certain condition. Using the Flick standard artifact as a quality evaluation tool noted a high precision increment in diameter and roundness form indication. This means a better transfer stability of CMM measurement quality could be significantly improved. Moreover, some error formulae of data sets have been postulated to correlate the diameter and roundness measurements within the application range. Uncertainty estimation from CMM and environmental temperature has been evaluated and confirmed the quality degree of confidence in the proposed verification method.

Part 4 discusses the important factors that affect the accuracy of roundness measurement and roundness determination method for the Talyrond metrology technique as an important technique in mechanical metrology in the automotive industry. Chapter 7 presents the effect of fitting algorithms on the roundness measurement accuracy to evaluate the ultra high-precision instrument, for roundness is an essential geometrical form in precision engineering, especially for rotating elements in aerospace and automotive industries. Roundness-measuring machine "Talyrond TR 73 HPR" is the one of major instrumentations in form and surface metrology. The software of roundness-measuring machine can contribute significantly to dimensional metrology. The characterization of the Talyrond machine software is very important to find an optimum solution for the roundness deviation. In this chapter, the impacts of Talyrond TR 73 HPR software parameters using region division method in roundness measurement have been studied. Ten metrological software parameters in this method have been estimated and discussed in detail. The measurement result revealed that the lowest deviation in roundness measurement has been investigated using MZ algorithm. The average of total roundness indicates low deviation by about 65.6% and 57.3% using 2CR- and Gaussian-filter, respectively, which proved the effectiveness within the application range. Furthermore, the experimental method was performed to establish reference data sets for circular, cylindrical, and spherical objects. The established data sets have been analyzed to help the software designer and the metrologist to satisfy the best measurement conditions in roundness nanometrology. Thus, the contribution of various measured results is estimated by way of region division. Moreover, experimental results show that the predicted an evaluation method is reliable and effective, especially in modern industry.

Part 5 of the book is very interesting and important due to the discussion of the specific new application processes of engineering metrology with respect to two different types of automotive engines to predicate their performance using the CMM technique (Chapters 8 and 9). The author shows two different application categories in coordinate metrology based on engine metrology before operation and fingerprint of engine wear after operation. Chapter 8 focuses on the use of engineering metrology to provide a new inspection method for new or overhauled engines. This chapter discusses CMM in automotive metrology, and the author points out that geometrical and dimensional metrology was used CMM in automotive engine. The author also shows the specific application category in coordinate metrology based on engine inspection before the operation. The main objective of this work is to develop an inspection program to demonstrate exploiting the advanced metrology devices in conducting precise and accurate dimensional and geometrical measurements on the crucial replaceable elements of water cold diesel engine during an overhaul. This new inspection method can be suitable for new engines. A correlation can also be made between the measured dimensional and geometrical deviations with the performance parameters of the overhauled diesel engine. These measurements include deviations in dimensions, form, location, and other geometrical features of the cylinder bores, valves, and valve seats. In addition, instantaneous peak compression pressures attained in cylinders representing the overhauled engine performance are also measured and correlated to the dimensional and geometrical measurements. Chapter 8 discusses another attractive and interesting new application for CMM to provide a new diagnosis metrology tool for automotive diesel engines. In this chapter, the author presented that the geometrical and dimensional measurements using CMM play an important role in automotive engine metrology. The author presents the new application of coordinate metrology based on engine inspection worn-out fingerprint of engine wear after operation. The main objective of this work is to obtain the accurate precise measurements of CMM machine in exploring and investigate the wear happening between contacting solid surfaces. For instance, excessive wear, if detected by the CMM measurements, in a cylinder bore of an internal combustion engine can dramatically affect its performance quality, sealing function, scheme of lubrication, and eventually its service life span. In such a case, the fingerprint would be the original design GD&T tolerances.

Chapter 10 discusses many specific applications using surface metrology in quality control methods automotive engines from the real industry's point of view. New applied technologies in the coating of crucial engine elements and its functions beside surface characterization are also presented.

Part 6, the last part (Chapter 11), presents some real developments, recommendations, and future work in this important area. This book is of interest to people in universities, research institutes related to accuracy and precision in automotive/equipment industries, and those who are involved metrology in production engineering and quality control.

To enjoy reading this book, it would be useful to go through a brief introduction to the following topics: engineering metrology, quality challenges in automotive engineering and metrology laboratory and automotive engineering. These topics include a brief presentation of some basic principles of scientific background related to the subject of the book for the non-specialist or general readers who wish to increase their knowledge.

1.2 Engineering Metrology

The word metrology is derived from two ancient Greek words (*metron*) means the meter and (*ology*) means the science. This is due to discrimination for the value of meter, because meter is the first unit of measure in the world and its history lies in Giza of ancient Egypt, once called "the mother of the world" and in Arabic "Misr Omo El-Donia." Historically, the first standard of length measurement was made from wood. It was called the "Egyptian royal cubit," based on the length of the Pharaoh's arm. The Egyptian royal cubit was the first standard tool for length measurement (see Fig. 1.1a). The cubit was developed by adding Egyptian Pharaonic numbers gradually, Fig. 1.1b. The cubit was used for the construction of the Egyptian pyramids (one of the Seven Wonders). After that it was manufactured from Egyptian black granite. By virtue of the cubit tool, the ancient Egyptian chariot was also made, as shown in Fig. 1.2.

(a)

(b)

No. 16   No. 15   No. 14

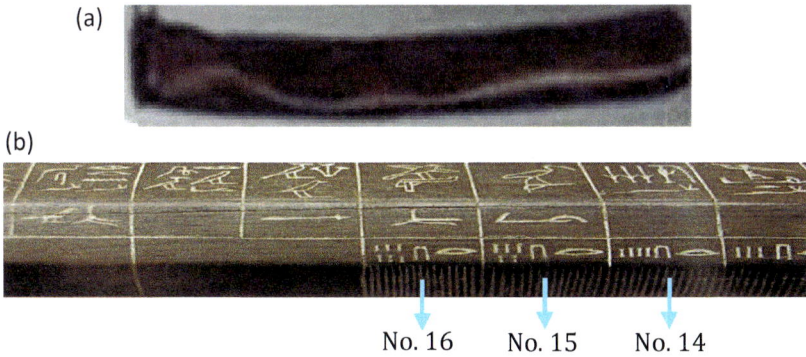

Figure 1.1   Ancient Egyptian royal cubit before 5k BC.

Figure 1.2   Ancient Egyptian chariot.

Then, the definition of the standard meter is even possible with different methods. In the middle of the 19th century, the need for a worldwide unified metric system became very important. In 1875, the first diplomatic conference on the meter was held in Paris; 17 governments signed the diplomatic treaty, which was named "Meter Convention." This convention introduced solutions for Technical Barriers for Trade agreement and interrelations between market, trade, conformity assessment, and accreditation (see Fig. 1.3). Typical measurement standards for subfields of metrology are shown in Fig. 1.4.

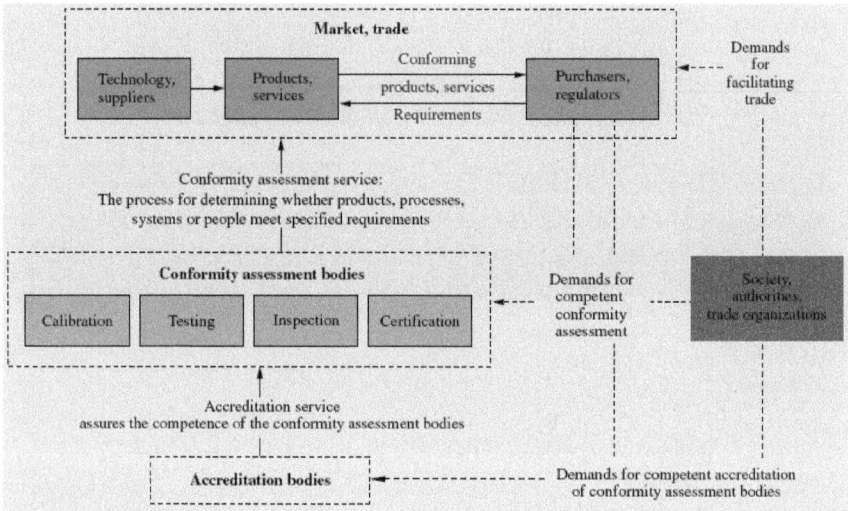

Figure 1.3 Block diagram of interrelations between market, trade, conformity assessment, and accreditation.

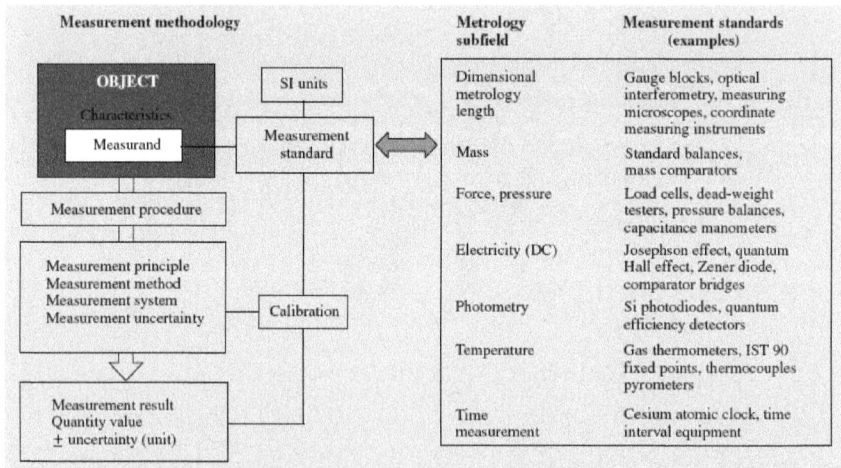

Figure 1.4 Block diagram of the measurement standards.

This was the beginning of the joint work based on a regular basis at the international level in metrology. Since 1983, the latest definition of the meter is "the path traveled by light in vacuum during a time interval of 1/299, 792, 458 of a second", where 299, 792, 458 is the speed of light in m/s.

Metrology is inseparable from the measurement for all things in our daily life. Everything in this world needs to be measured. If you cannot measure, you cannot make. We could say: good performance from an automobile cannot be obtained without engineering metrology. Thus, we can also say "metrology is the constitution of the sustainable economical advanced life." Metrology science includes all theoretical and practical aspects of measurement. Metrology is the measurement science with the level of uncertainty and its application. Metrology deals with

- measurement theory;
- measurement units and their physical realization;
- measurement processes, procedures and methods;
- measuring instruments, verification, validation, and character-izations

Metrology may be divided into those four basic activities due to the overlap between the following three categories:

- definition of international accepted SI units.
- realization of these units of measurement in practice.
- application of traceability chains of measurements made in conformance to reference standards.

Metrology is a very broad field, especially in automotive engines. Engineering metrology consists of three major types of activities:

- scientific metrology
- industrial metrology
- legal metrology

Scientific metrology depends on the applied research conducted in metrology to establish quantity systems and units of measurement; develop or create new measurement methods; realize measurement standards and transfer the traceability to users in society in order to achieve their desired benefit. Industrial or applied metrology concerns the application of measurement science to the manufacturing and production processes, the main motivation for writing this book. This ensures using the suitability of measurement techniques, their calibration, and quality control of measurements. Although the emphasis is on measurements themselves (especially for new measurement methods in scientific metrology), traceability of the calibration of the measurement devices is important to ensure confidence in the measurements. Legal metrology interests in all different activities which result from statutory requirements, units of measurement, measuring instruments and tools, methods of measurement and which are performed by competent bodies for all transactions in our daily lives. In conclusion, the author emphasizes that the engineering metrology is the essential foundation to achieve an advanced quality system.

1.3 Quality Challenges in Automotive Engineering

Quality and efficiency for many industrial processes and operations depend on the metrological integrity of instruments with regard to accuracy and precision. Quality system certification of products often requires accurate and precise test and inspection equipment to have international traceability. Nowadays, automotive industry must deliver high levels of innovation, flexibility, lowest cost with the highest quality using accurate and precise measurements. Metrology represents both significant opportunity for innovation in light of increased complexity in current technology. Automotive engineering organizations using accurate and precise metrology techniques to integrate the capabilities need to accelerate innovation and improvement more efficiently. Due to metrology's success in finding superior solutions, the leading global enterprises in the automotive engineering industry continue to use the true sense of metrology for solving their most complex and critical development challenges with new regulations and ISO standards to improve product quality. Thus, metrology must achieve high quality while controlling costs and time. The customer's expectations of high product quality influences the performance, safety, reliability, and profitability. However, today's highly competitive and global automotive market demands high performance, cost efficiency, and the ability to move new products to market rapidly. Metrology has a proven record of delivering solutions to manage complex changes and enable collaboration, traceability, and transparency from product inception through realization even to the good performance at use. The quality of automotive engine performance, operations cost, time consumption, and also the company image are really based on the suitable use of advanced accurate and precise metrology techniques (see Fig. 1.5).

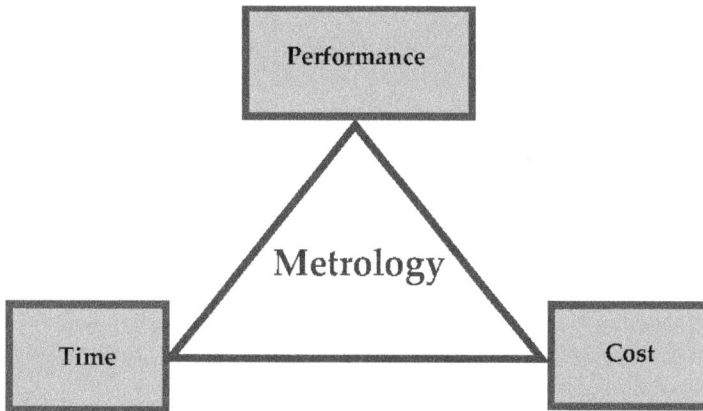

Figure 1.5   The core role of metrology.

With regard to quality challenges, I believe that the high engine performance corresponds to high quality of inspection process. It focuses on five basic strategies that significantly affect inspection and product quality [1]:

- full traceability of metrology across disciplines, artifacts, teams and product variants and errors for precise impact analysis.
- process management with improvement through support of the ISO standards.
- requirement-driven development—metrology integrity provides a complete requirement management solution, including requirement change and technical configuration, requirement traceability across all disciplines, authoring right, and requirement-based test management, etc.
- development asset reuse using experiences and reference, not copy and paste, the tracking of issues and defects across product lines

- release readiness visibility based on a single source of truth for development (R&D lab according to metrology lab) to empower accurate decisions and eliminate bottlenecks across disciplines

1.4 Metrology Laboratory

Many coordinate metrology techniques such as CMM and Talyrond machines are available for automotive engine metrology. Measurement procedures and testing and inspection methods have to be available, reliable, and traceable to perform high-quality requirements according to ISO.

A metrology laboratory that may be used by a manufacturer comes in two basic functions: calibration and testing. Testing and calibration labs have to receive accreditation under the ISO/IEC 17025 standards. However, ISO/IEC 17025 has its own unique set of technical and management requirements as well. Calibration lab may be internal or external. Calibrations are often performed in external accredited laboratories or in national metrology institutes (NMIs) such as NIST in the USA, KRISS in Korea, PTB in Germany, NPL in the UK, and NIS in Egypt. The testing lab may also be differentiated further into tools, artifacts, materials and other categories and may be internal or external. The global automotive industrial companies may use a testing laboratory for any of the following reasons [2]:

- to perform tests to ensure that a product is ready to be introduced to international markets
- to meet regulatory or ISO requirements and ensure perform properly
- to have certified products by organizations according to requested
- to obtain a particular listing such as the Evaluation Services report from the International Code Council for products

In conclusion, the author expects the increased use of technologies of modern engineering metrology to accomplish high-quality auto engines, which can compete in the automotive community all around the world.

1.5 Automotive Engineering

Automotive engineering, along with aerospace engineering, marine engineering, and locomotive engineering, is one branch called "vehicle engineering." The word "automotive" is used mainly in automotive engineering, which deals with the design, manufacturing, diagnostic, service of maintenance or repair and operation of automobiles such as passenger cars. The word "automobile" is a combination of old two Greek words, "*autos*" meaning self and "*mobilis*" meaning moving. So, an automobile means anything that moves on its own. An automotive engine is a unique unit and important in automobiles. An automotive engine incorporates the elements of mechanical, electrical, electronic, and management software system as applied to the design, manufacture, and operation of automobiles, buses, trucks, ships, and airplanes. Figure 1.6 shows the relation between automotive engineering and metrology. An automobile comprises many different systems such as engine, transmission, suspension, and braking system. No other system in an automobile is more important than the source of power in the automotive engine. Figure 1.7 illustrates an automotive engine. The main purpose of the engine in the automobile is to generate power by transforming chemical energy into kinetic energy.

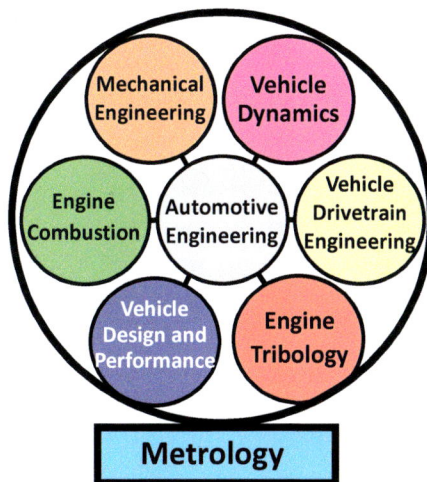

Figure 1.6 The role of metrology in automotive engineering.

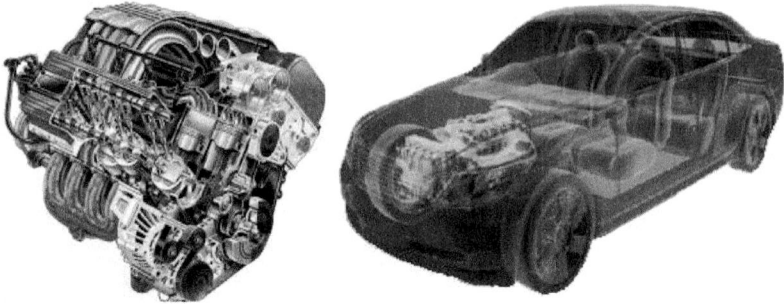

Figure 1.7 Automotive engine is the basic unit.

1.5.1 Types of Automotive Engines

Automotive engines are mainly heat engines or electric engines. The heat engine is classified into internal or external combustion engine or steam engine. Most automotive engines are internal combustion engines (ICEs). The ICEs are engines in which the internal combustion of a fuel–air mixture takes place in a combustion chamber. In an ICE, the expansion of the high temperature–induced high-pressure gases by combustion applies direct force on the piston in the engine cylinder. The force applied to the piston moves the component over a stroke, transforming chemical energy of the air–fuel mixture to useful mechanical energy to transmission system and then to wheels. ICEs are also divided into two main types: four-stroke engines and two-stroke engines. Automotive engines are also distinguished by the relative dimensions placement and number of cylinders. Figure 1.8 illustrates the common four-stroke cycles of diesel engines.

Automotive engines are also classified into cylinders alignments: in-line engines, V-engines, flat engines, rotary engines, etc. Then, they are classified on the basis of the value location present in the cylinder head or block, cylinder capacity, cooling system, function, and so on. Today four-stroke overhead valves are commonly used in an automobile. The fuel used in automotive engines can be of three main types: diesel, gasoline, and natural gas.

Diesel engines are the very low cost fuel than gasoline. They also have certain advantages with respect to fuel consumption costs and long service lifespan. Diesel engines are used mainly in trucks and buses, microbuses, and passenger cars. Gasoline engines are superior because of their design, lower cost of production, power, good starting, and better control on the exhaust smoke. However, modern high-speed diesel engines have certain properties that brings them almost at par with gasoline engines. Diesel engines nowadays have begun to find applications in various types of automotives. The types of modern diesel engines according to cylinder alignment are shown in Fig. 1.9a–c. The type of straight or in-line diesel engine has cylinders arranged one after another in one straight line. These engines are considerably easy to build. They also have ultimately low production cost and maintenance cost. They are also very lightweight and therefore are preferred in front-wheel-drive cars. They are extremely fuel efficient than v-type engines (see Fig. 1.9a). V-type diesel engines have two sets of cylinders placed at about 90°. They have various advantages such as shorter length with the same efficiency, greater rigidity, heavier crankshaft, and simple attractive profile (see Fig. 1.9b). Horizontal, boxer, or flat diesel engines have a low center of gravity than other type of engines. The automobiles that use this type of engines have certain benefits such as better stability and control. These engines are wider. Thus, it is difficult to install them in front-engine vehicles (see Fig. 1.9c).

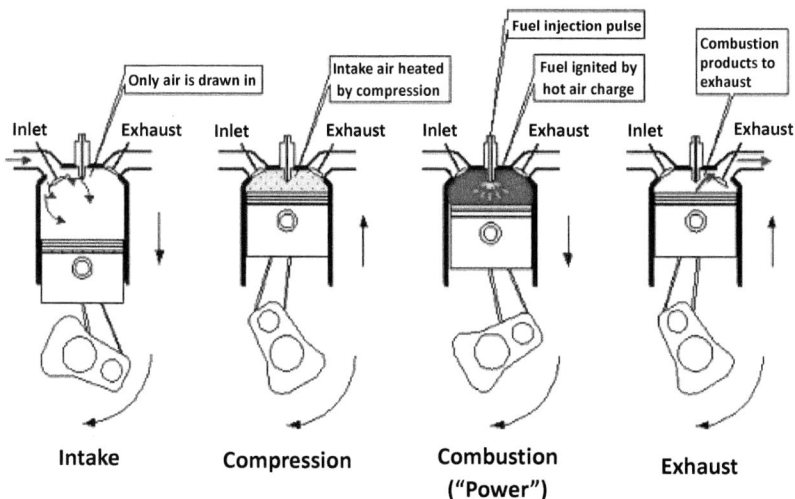

Figure 1.8   Four-stroke cycles of diesel engines (compression-ignition) [3].

Figure 1.9   Automotive diesel engines: (a) Straight line-type engine.

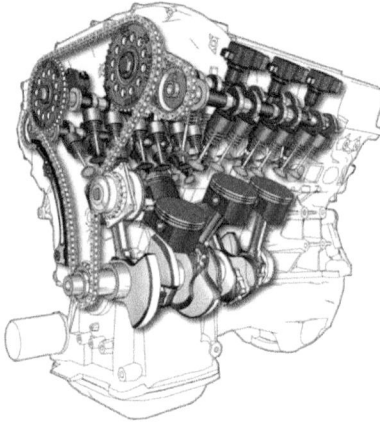

Figure 1.9   (b) V-type engine.

visit at www.blaze of automotive.wordpress.com

Figure 1.9   (c) Flat or horizontal-type engine.

1.5.2   Performance of Automotive Engine

Automotive performance is a measurable and testable value of engine power and torque performed in various conditions. Automotive performance is a basic function of the maximum compression pressure generated in engine cylinders. Accordingly, it is a function of the machining accuracy and the degree of finish of the inner surfaces of the engine parts and cylinders. The balance in an internal combustion engine is presented in Fig. 1.10. Figure 1.11 shows the dynamic behaviors of automotive diesel engine performance at different engine speeds. Automotive performance can also reflect the amount of comfort, steering, and control in inclement weather. Performance can be considered in a wide variety of experimental tasks, but it is generally associated with the vehicle acceleration, how a vehicle can come to a complete stop from a set speed and distance, how much power it can generate with stability, cornering critical speed, brake fade, etc.

Many controlling factors affect the performance of an automotive engine:

- compression ratio
- air/fuel ratio
- combustion duration
- combustion timing
- pumping loss
- surface finish of tangent parts
- mechanical friction loss

Better fuel economy by reducing fuel consumption is another important target in automotive engines. Serious pollutants generated in the engine exhaust gases are nitrogen oxides (NO_x), carbon oxides (CO_x), hydrocarbons (HC), and particulate matter (PM). The PM is referred to as small particles leaving the engine, mainly constituted by carbonaceous material. These fine particles (size less than 5 μm) can be inhaled by human lungs and are responsible for some respiratory and cardiovascular diseases [5]. Diesel engines can provide better fuel economy and produce lower CO_2 emissions than conventional gasoline engines. However, the NO_x gas components in the lean (oxidizing) exhausts from diesel engines cannot be efficiently removed with the classical three-way catalyst (TWC) under operating conditions according to the oxygen rate in the exhaust gas [6].

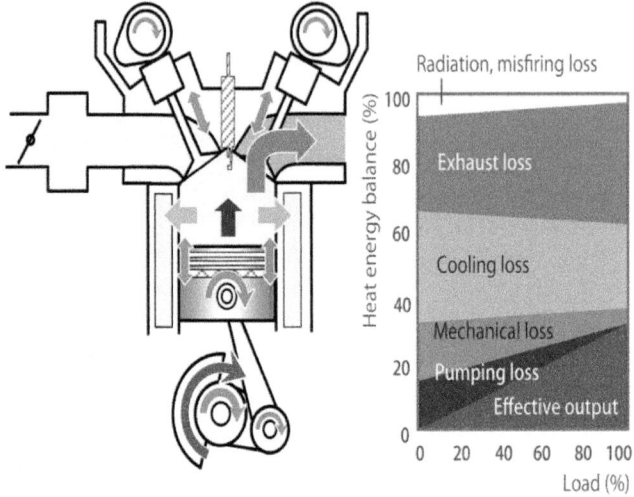

Figure 1.10 The balance in an internal combustion engine [4].

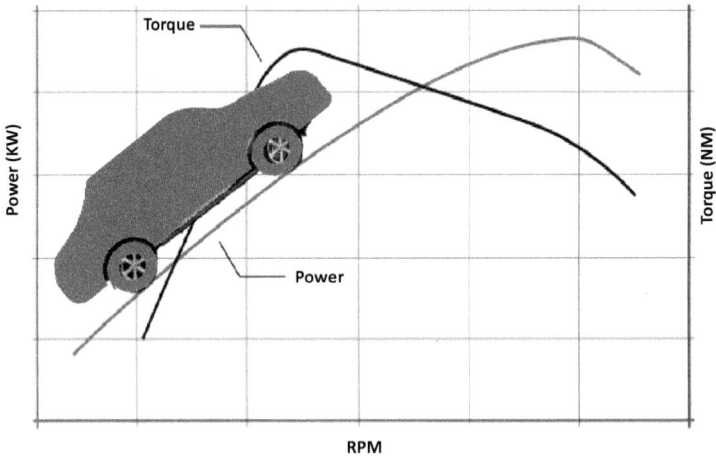

Figure 1.11 The performance of an automotive diesel engine.

Eventually, it can be concluded that the automotive performance is directly related to the engine performance. Thus, the automotive engine performance is directly influenced by the surface finish of engine cylinders.

1.6 Conclusion

It has become clear that there is a close relationship between automotive engineering and engineering metrology. Moreover, the discussion in this chapter leads to the following question: Which one deserves to be thanked first: the science of metrology or the automotive science, and why?

References

1. Quality through traceability, reuse and visibility. On website at: http://www.mks.com/solutions/by-industry/automotive/quality.

2. Hershal C. Brewer, Working with an accredited Calibration or testing laboratory has many benefits, July 3, 2014. On Website at: http://www.qualitydigest.com/may05/articles/05_article.shtml.

3. Why choose a diesel engine, August 23, 2013. On website at: http://automobileandamericanlife.blogspot.com/2013_08_01_ archive.html.

4. Ideal Engine, Green Car Congress, August 10, 2011. Website at: http://www.greencarcongress.com/2011/08/skyactiv-20110804. html.

5. G. Mazzarella, F. Ferraraccio, M. V. Prati, S. Annunziata, A. Bianco, A. Mezzogiorno, G. Liguori, I. F. Angelillo, M. Cazzola: Effects of Diesel Exhaust Particles on Human Lung Epithelial Cells: An in vitro Study, *Respiratory Medicine Journal*, vol. 101, no. 6, pp.1155–1162, 2007.

6. B. Pereda-Ayo and J. R. Gonzalez-Velasco, NO_x Storage and Reduction for Diesel Engine Exhaust After-treatment, Dept. of Chemical Engineering, Faculty of Science and Technology, University of the Basque Country UPV/EHU, Bilbao, Spain, 2013.

PART 2
ADVANCED METROLOGY TECHNIQUES

Chapter 2

Advanced Measurement Techniques in Surface Metrology

Surface metrology has become very important in many branches of science and industry. The study of dimensional and surface nanometrology is becoming more commonplace in many applications and research environments as well as data treatments dealing with standardized rules. Therefore, surface characterization using advanced accurate and precise nanomeasuring techniques is important scientific tools especially in the automotive engineering, tribology, biotechnology, and criminology. Because of this diversification, there are more advanced metrology techniques using stylus, optical, and non-optical methods used for analyzing the surface characteristics, where each technique has its own specific applications [1].

The ISO technical committee TC-213 in the field of dimensional and geometrical product specifications and verification formed a working group WG-16 to address standardization of areal (3D) surface texture and measurement methods, in addition to review existing standards on traditional profiling (2D) methods including characteristics of instruments. In 2007, the project of this working group was to develop standards for three basic methods of areal surface texture measurements [2–4]. A line-profiling method used a high-resolution probe to sense the peaks and valleys of the surface topography and produce a quantitative profile $Z(X)$ of surface. Areal topography methods extend the line-profiling method into 3D, usually by restoring a series of

Automotive Engine Metrology
Salah H. R. Ali
Copyright © 2017 Pan Stanford Publishing Pte. Ltd.
ISBN 978-981-4669-52-8 (Hardcover), 978-1-315-36484-1 (eBook)
www.panstanford.com

parallel pattern profiles or by some quantitative topographic imaging process. It is important to note that some types of areal profiling methods can sense Z-height as a function of both the X-and Y-coordinates, whereas others can display topographic images with a series of parallel $Z(X)$ profiles whose relative heights along the Y-direction may be somewhat arbitrary.

In this chapter, novel techniques used such as coordinate measuring machine (CMM), roundness testers, roughness measurement instrument, white-light interferometer (WLI), confocal optical microscopy, digital holography, scanning probe microscopy (SPM), and computed tomography (CT) methods are investigated. One of the major challenges when moving to smaller measurements is selecting the suitable metrology technique for the desired measurement. Standardization processes will be probably increased in terms of methods of measurements and data treatments for specific applications due to the externalization and diversification of products. These different advanced techniques have currently advantages meeting the samples specifications and required properties. Therefore, it is important to be aware of how techniques can affect the measured parameters according to specific accurate and precise strategy of measurement. The new applications of dimensional and surface nanometrology in materials science, automotive industry, tissue engineering, and banknote paper will be considered. Moreover, future directions under development will be presented and discussed scientifically in order to introduce proposed solutions for the issues that need to be addressed in the area of interest.

2.1 Advanced Measuring Techniques

The 3D surface metrology techniques have been rapidly developed in the last decade attributed to the advanced computational software technology. Mechanical contact stylus, optical and non-optical measurement techniques achieved significant progress in many different applications.

2.1.1 Mechanical Contact Stylus Techniques

The tactile stylus measurement techniques as a coordinate metrology and roundness facilities are powerful tools in modern engineering industries [5–7]. In these techniques, the stylus profiler senses the surface height through mechanical contact, while the stylus traverses the peaks and valleys of the surface with a small contacting force. The vertical motion of the stylus is converted to an electrical signal by a transducer, which represents the surface profile $Z(X)$ or areal topography image $Z(X, Y)$. Therefore, the stylus measurement technique is directly sensitive to surface height with little interference. Two disadvantages of the stylus instruments, however, are that the stylus may damage the surface depending on the hardness of the scanned surface relative to the stylus normal force and the stylus tip size [2]. The most important stylus metrology techniques such as coordinate measuring machine (CMM), roundness Talyrond, and surface roughness devices are the three major developed technologies.

2.1.1.1   CMM coordinate technique

Advances in software during the 1980s allowed CMMs to have error corrections and provided geometric computations [8, 9]. Now, the CMM technique is one challenge for advanced coordinate metrology in modern engineering applications. The basic function of CMM is to measure the actual geometrical shape of object compared to desired shape and evaluating the collected data using metrological aspects of size, form, location, and orientation [10]. The actual shape is obtained by probing the surface of the object at definite measuring points. Figure 2.1 illustrates four common different types of developed CMMs. Additionally, the advantage of developed CMM techniques is to convert the data of the measured object into 3D image suitable for other CAD/CAM systems.

Improving the accuracy of CMM measurements is another important issue according to ISO Standards. ISO-10360 deals with verifying the performance assessment as specific value of permissible errors for advanced CMM technique [11, 12].

However, there are two methods used to increasing the accuracy of CMMs [13]. The first is based on research and development of design parameters influencing the CMM errors. The most important parameters affecting the CMM errors are the probing system, digital computer with software and environmental conditions [14–25]. The second development is depending on calibration methods using different devices such as ball plate artifact, hole-plate artifact, and laser tracker.

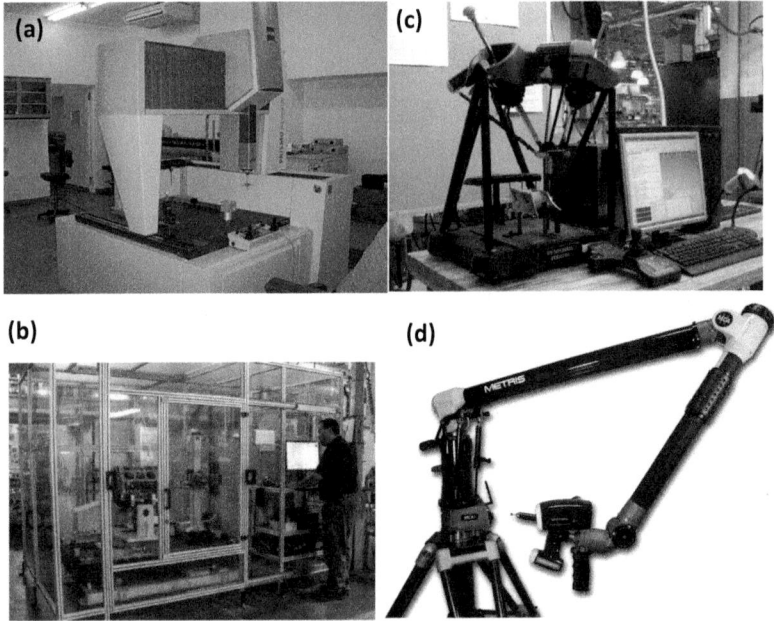

Figure 2.1   New types of CMMs: (a) CMM Prismo type with ±0.1 μm resolution, (b) Automotive engine-block quality station using CMM shown in (a), (c) Small CMM of automotive spare parts, (d) Portable arm CMM.

The new-generation advanced coordinate metrology provides ultra-precision CMM with large measurement volume 400 × 400 × 100 mm^3 (Fig. 2.2). The new Isara 400 CMM is the latest development by IBS Precision Engineering in the Netherlands. This Isara 400 CMM machine enables coordinate metrology of large and complex parts with nanometer level of uncertainty. The expected length measurement uncertainty is 45 nm, while the full-stroke 3D measuring uncertainty is expected to be 100 nm at a 95% confidence level.

Tactile probes, such as the presented Triskelion ultra precision touch probe, as well as other possible (optical) probe systems, can be used to perform scanning at discrete points of object. The presented Isara machine provides a technology that can be adapted and optimized for specific user requirements. While in the future, there is another new trend to produce a new CMM machine with multiple sensors (tactile and optical) in a cooperation research. It will be used in many applications with relative high level of accuracy [27]. On the other hand, the new multi-sensor CMM has been designed [27]. It is equipped with both a contact probe and an optical sensor. The profile resolution measuring range is 2.5 nm and MPE is limited to 250 nm, it is an ideal solution for applications in the plastic industry, medicine, and automotive technology and in precision mechanics when a large number of components have to be measured in the short time with at high accuracy.

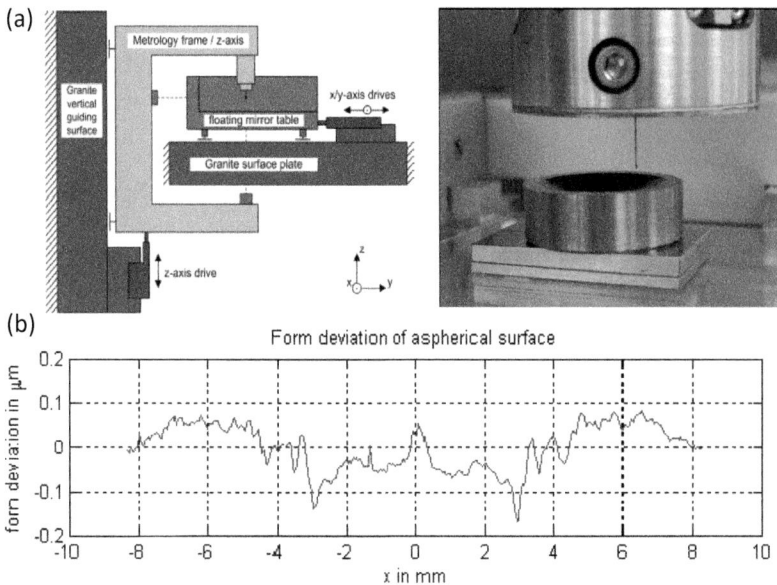

Figure 2.2 New Isara 400 CMM [26]. (a) Concept of new CMM; (b) measurement using the Triskelion probe.

The future trends in advanced CMM have been discussed. A novel type compact five-coordinate measuring machine with laser and CCD compound probe have been designed and built up as presented in Ref. [28]. Mechanical structure and measurement model are introduced. Five motion axes include three translational axes *XYZ* and two rotational directions AB, as shown in Fig. 2.3. The laser displacement sensor and CCD camera compose the compound probe that can rotate around A and B axes while moving along *Z*-axis. The tested workpiece lies on the worktable that can travel along the *X*- and *Y*-axes in horizontal plane.

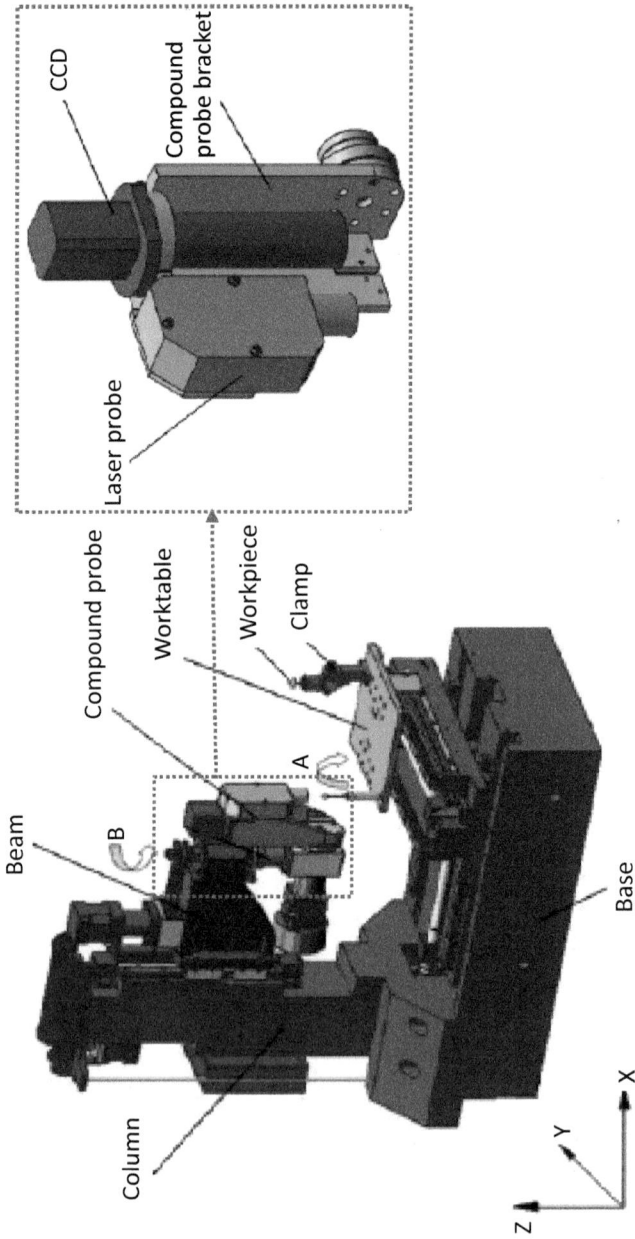

Figure 2.3 The overall mechanical structure of five-coordinate measuring machine [28].

The step motors with subdivision function are applied to drive each axis. Among them, the motor for A direction has contracting brake device which will lock the motor mainshaft and prevent collision of the probe and worktable in case of accidental power off. The high-precision ball screws are applied to drive X- and Y-axes. Thus, the related experiments illustrate the characteristics of update CMM machine as well as the feasibility and validity of mentioned methods. Eventually, this area usually needs more dynamic analysis to understand probe response according to the design and construction of CMMs, especially new hybrid CMM machines [11, 28, 29].

2.1.1.2 Roundness instrument

The profile form of the roundness is of primary significance and is an important aspect of engineering surface for cylindrical features, especially in the modern investigations and quality control problems. The development of software analysis tools and their validation is also another major challenge facing the industry. Advanced mathematical techniques are incorporated into such software systems to provide further reference algorithms [30–37]. In this section, two directions of research experiences are reviewed [32–35]. Moreover, new ideas in this area may promote new techniques in the future [36, 37].

In Russia, Bogomolov et al. [32] modified the rotating table (Talyrond-200) and rotating gauge (Talyrond-73) machines to be automated. These developed systems allowed to apply modern computer-based measurements and analysis of roundness and waviness parameters. Each of the measurement systems consists of analogue roundness machine, analogue-digital converter (ADC/DAC) board, connection cables, and PC with specific installed software. An example of the 2D roundness measurement results of the component with parametric analysis in software is presented in Fig. 2.4a. Waviness analysis was provided by proposed measurement systems, which is performed by means of reverting the roundness/waviness filter. Multiple profile 3D analysis of the face side waviness by the measurement systems were also provided as shown in Fig. 2.4b. The proposed measurement system provides automated high-precision measurements and complex PC-based analysis of roundness and waviness. In this research, inexpensive and powerful automation of both rotating table and rotating gauge roundness machines was performed with the financial support of the Russian federal target program "Academic and teaching staff of innovative Russia in 2009–2013".

In United Kingdom, from year 2000 until now, Taylor Hobson entered a new phase of innovative product design using advanced software [33–35, 37]. Roundness measurements on an automotive fuel injector have been detected using the new Talyrond-365. The 2D graph represents the radial variation output form of the surface measurement for injector as a polar profile. The circumference measurement shows deviation form of the surface roughness and the presence of a scratch that can cause leakage. A scratch is more likely to cause component failure if it is aligned with the component axis. The pass/fail criteria can be programmed and data exported to software of the averaged measurement data.

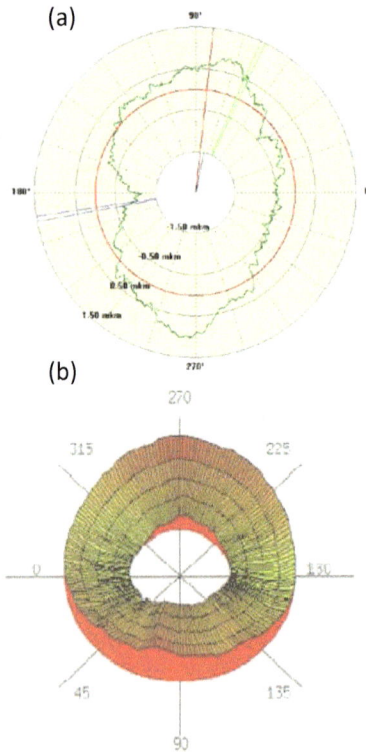

Figure 2.4 Sample results of developed roundness measurement system in Russia [32]: (a) 2D roundness measurement; (b) 3D waviness measurement.

The analysis of structured surfaces, those with repetitive patterns, requires entirely new methods and techniques [38]. Surface-wavelength-based filtering and average parameters will be less applicable in such instance, while discrete feature geometry-based parameters and their statistics over the surface appear to be more relevant. The new wave in advanced 3D roundness metrology, the design of new ultra-precision Talyrond using ultra software has been presented [35, 38]. Figure 2.5 shows the 3D results obtained from Talyrond-395 [35]. These techniques work as an automatic centre and leveling spindle for measuring the workpieces. These work dynamically and the anti-vibration mounts combined with the environmental enclosure provide excellent isolation from external influence allowing confident measurement to the nanometer level. Talyrond-385 and -395 techniques perform fully automated roundness geometry system incorporating micro-ultra software.

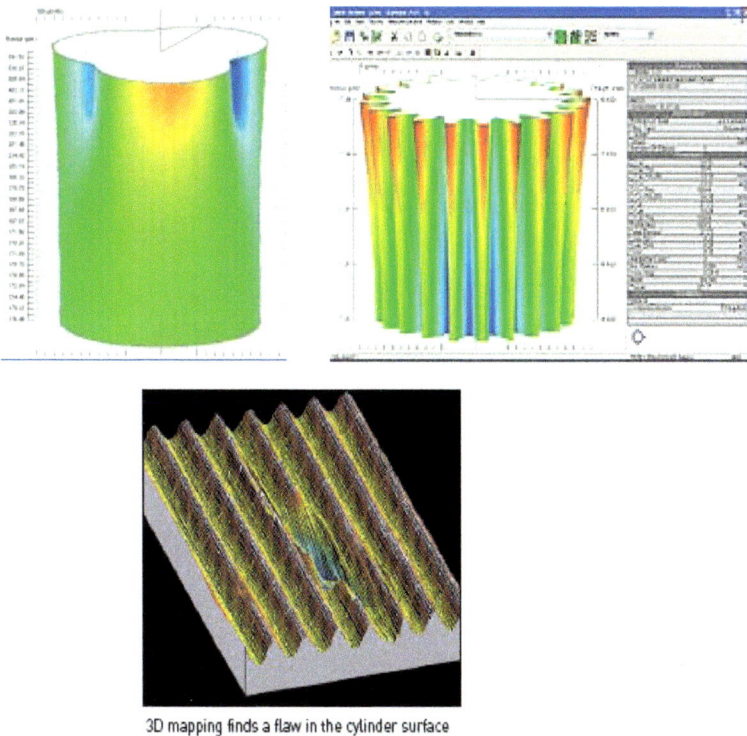

3D mapping finds a flaw in the cylinder surface

Figure 2.5   Results obtained from Talyrond-395 [35].

On the other hand, new ideas are suggested for new roundness measurement techniques in the future [36, 37]. For multi-probes technique, Gao et al. proposed a combined three-point method for on-machine roundness measurements [36]. The roundness measurement system uses three capacitance-type displacement probes as shown in Fig. 2.6. These three probes are fixed around the object to detect the workpiece profile of roundness form, which can be extracted. Spindle error components can be filtered. The combined three-point method not only cancels the effect of the spindle error but also measures roundness profile, including the stepwise variation. The feasibility of this proposed method for roundness measurement has been confirmed using PC simulations.

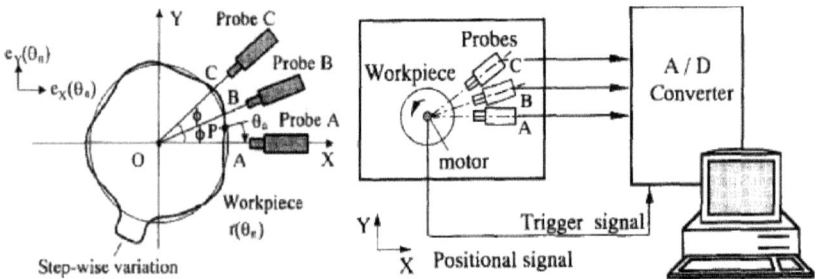

Figure 2.6   Principle of the combined three-point method for roundness measurement [36].

Another idea of non-contact probe has been proposed to measure the diameter and the roundness of turned objects [37]. This research discusses the implementation of roundness measurements into the initial probe and its performance. The principle of roundness measurements is based on the fact that the light intensity varies with respect to the displacement from the reflective object. The displacement between the probe and the rotating workpiece may be varying according to the out of roundness (OOR) of the object, the error of the rotational system, and the radial throw of the object. The variation of the displacement causes variation of the light intensities reflected from the rotating workpiece. The OOR of the rotating workpiece can be determined by analyzing the variation of the light intensities. The developed probe configured for diameter measurement can be easily reconfigured to measure the OOR of the rotating workpiece by blocking only one laser beam. Figure 2.7 shows the probe for roundness measurement. All components in the probe configured for roundness measurement are the same as those configured for radius measurement. In diameter measurement, two laser beams are introduced. In roundness measurement, one laser beam is blocked before reaching the focusing lens. A blockage can be placed between the beam splitter and the focusing lens. The laser beam is directed to the object surface. The receiving lens collects the scattered light and focuses it to the photodiode. The intensities of the scattered light are converted to an electrical signal by the photodiode. The electrical signal from the photodiode is sent to an amplifier circuit to eliminate undesired noise and prepare the signal to match the input requirements of the data acquisition (DAQ) system.

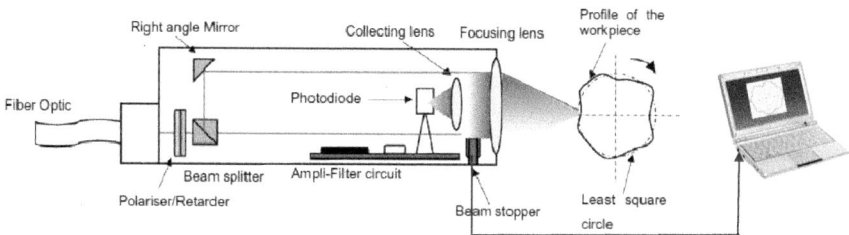

Figure 2.7   The probe configuration for roundness measurement [36].

The signal from the amplifier circuit converts to the PC with LabView environment and is processed to extract the OOR of the rotating workpiece. The result shows the extensive roundness tests have been determined to validate the performance of the probe. The measurements made using the laser-based probe are averages of at least 10 repeated measurements. The out-of-round results compared with the results obtained from Taylor Hobson. The probe performance gives a maximum error of 0.5 μm with an uncertainty of 1 μm for roundness measurements. This concluded that geometrical form of the roundness measurement can be adapted from the radius measurement form. This is a potential capability to add measurement features to the existing probe, e.g., surface roughness. The signal obtained from scattered lights using the roundness configuration includes the roundness information and the roughness information.

2.1.1.3 Roughness measurement technique

A profilometer technique was first constructed by Abbott and Firestone in 1933 [39]. It is known that modern software allows computing of approximately 300 parameters of roughness profile and dozens of topography parameters. Roughness of any surface can be measured up to 200 mm length and 100 mm width with the deviation of guide equal to fractions of micrometers, and further software support of accuracy can be applied.

In this case, the slide can be measured by a laser interferometer and its errors can be collected in the microprocessor system and used for the correction of indication. Besides, the measuring instruments often offer simultaneous measurement of roughness and outline with greater range—even above 2 mm with 0.6 nm resolution. The interesting element of this device is a probe— magnetically fixed, which prevents damage of any impact or overload applied to the part. The diamond needle is separated from the body on three-point magnetic holder. Additionally, the probe is equipped with an amplitude modulation transmitter and a receiver, which is used for communication with a central processing unit as presented in detail in Refs. [40, 41] (Fig. 2.8).

Figure 2.8   Two modulation of amplitude in roughness measurement [40].

Next possible solution is the application of the measurement probe with bidirectional pressure force—up and down. It allows to measure, using the same probe, the roughness on the upper and lower surface of a hole, and in conjunction with an incremental linear encoder in a measuring column allows measurement of small inside diameters along a vertical axis. It is especially useful in the automobile industry, production of pumps and injectors, where without fear of probe damage, CNC procedures to measure roughness, waviness, and outline in small holes can be used. In another configuration, the application of developed cantilever-type tactile sensor for fast and nondestructive form and roughness measurements has been presented. Fuel injector nozzle comprising spray holes of 170 to 110 μm in diameter were characterized at different scanning speeds and probing force. Thus, results of that work proved the potential of the novel sensor for in-process metrology during the manufacturing of injector nozzles [42, 43]. The tactile sensor has been realized as an extremely slender silicon cantilever integrated with probing tip. When the tip is brought into contact with workpiece during test and moves along the surface, the supporting cantilever is deflected which can be monitored via an integrated piezoresistive strain gauge. The typical profile measured inside the spray hole, obtained by scanning a tactile cantilever sensor tip, which is composed of distinct regions has been presented [43].

Measurements based on a stylus profilometer in 3D surface topography are time consuming, which is a significant limitation. A possibility of overcoming this inconvenience is spiral sampling [44], as shown in Fig. 2.9 [45]. Irrespective of the contact devices, constructions based on optical phenomena are being developed. Therefore, the optical methods have been described in-depth in the scientific literature [45–49]. New solutions have introduced CCD (charge-coupled device) lines and arrays to detect the light signal as used in the light scattering methods. These techniques can be used successfully in roughness measurements in preventive inspection, and their vertical measuring range reaches one micrometer [50, 51]. Modern interferometers, used in roughness measurements, are systems applying white light. The most popular interferometric measurement techniques are phase-shifting interferometry (PSI), vertical scanning interferometry

(VSI) and enhanced vertical scanning interferometry (EVSI). PSI uses a monochromatic light source and generally is applied to the analysis of very smooth surfaces, because this method is characterized by sub-nanometer resolution. On the other hand, it suffers from phase-ambiguity problems, which limits PSI usability to a surface discontinuity not higher than $\lambda/4$, where λ is the wavelength of the light used. Besides, the monochromatic light source limits using PSI to ranges where continuous fringes can be obtained. In order to overcome this difficulty a new technique called multiple wavelength interferometry (MWI) has been developed which has extended high-difference limitation successfully. In this technique, two wavelengths are selected, which allow the user to increase the dynamic range and at the same time to keep the resolution constant. Further increasing of the dynamic range is possible when white light is applied (VSI). Then the continuity of fringes is not so crucial also more important is finding a focus. The principle of VSI system is shown in Fig. 2.10.

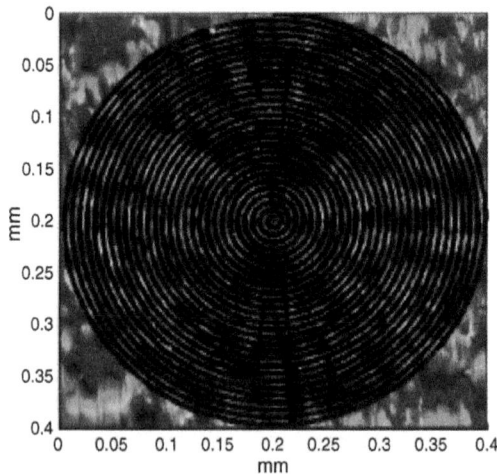

Figure 2.9   Application of spiral sampling [45].

Unfortunately, the resolution of VSI is in the nanometer range, not in fractions of nanometers. The advantages of PSI and VSI are combined in the EVSI technique, also called white light interferometery with phase-shifting [53]. First, every pixel was found in the optimal position of the objective for which this point of the surface is in the focus of the optical system, so in this position the distribution of intensity of the interference signal has a maximum. Second, to analyze the intensity around the point of focus, a PSI technique is used [43, 53].

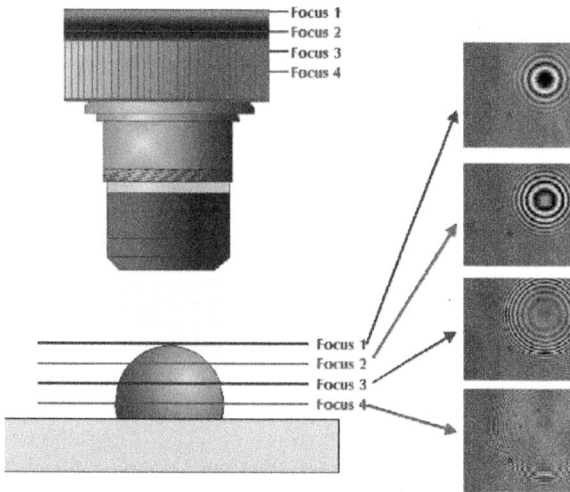

Figure 2.10   Principle of a VIS [52].

2.1.2   Optical Measurement Techniques

Light is at once the most sensitive and the gentle probe of measurement. It is easy to generate using light-emitting diodes or lasers, and to detect using ultrasensitive photodetectors. Light has become an indispensable nanometrology tool for surface characterization. For these reasons, a number of optical techniques have been developed for line profiling and areal topography. Nearly all of provide spatial resolution approaching the diffraction these techniques limit of optical microscopy. Optical methods have the advantage that they are non-contacting and hence, non-destructive tests. Optical methods based on imaging and microscopy also have a higher speed than contacting techniques which rely on mechanical scanning of a contacting probe. However, optical methods are sensitive to a number of surface qualities besides the surface height. These include optical constants, surface slopes, fine surface features that cause diffraction, and deep valleys in which multiple scattering may occur. In addition, scattering from tested surfaces within the optical system produces stray light in the system that can affect the accuracy of an optical profiling method. More sensitive methods, such as phase shifting interferometric (PSI) microscopy, have vertical resolution of the order of 0.1 nm [54, 55]. By contrast, WLI microscopy and confocal microscopy have a large vertical range, which is limited mainly by the range of the motion stage used to drive the vertical scan of the instrument and is often of the order of one millimeter.

2.1.2.1   White-light interference microscopy

Interferometers and microscopes are combined in interferometric microscopy. Through this combination, very good resolution and significant vertical range can be obtained. Interferometry as a measurement tool is certainly not new but combining old interferometry techniques with modern electronics, computers, and software has produced extremely powerful measurement tools [56–62]. The interferometer is responsible for scanning on a nanoscale, and the microscopy head is displaced on a microscale giving a vertical range even 1 mm. Descriptions of the operation exceeding and construction of scientific interferometric microscopes can be found in [63, 64]. Typically, there are two different techniques commonly used in phase shifting interferometry (PSI) and scanning WLI. WLI microscopy uses a broadband light source [65–67]. The optical system focuses the light through a microscope objective onto a surface. A principle schematic of a scanning white light system [68–71] is shown in Fig. 2.11. The upper beam splitter directs light from the light-source towards the objective lens. The lower beam splitter in the objective lens splits the light into two separate beams. One laser beam is directed to the surface of object and the other beam is reflected beam directed to a smooth reference mirror. When the reflected beams are recombined, interference fringes are produced around the equal path condition for the two beams. This equal path condition can be detected for each local area of the surface corresponding to each pixel of the camera detector. Scanning the surface vertically with respect to the microscope and detecting the optimum equal path condition at every pixel in the camera results in a topographic image.

A white light system is used rather than a monochromatic light system because it has a shorter coherence length that will give greater accuracy. Different techniques are used to control the movement of the interferometer and to calculate the surface parameters. The accuracy and precision repeatability of the scanning white light measurement are depending on the control of the scanning mechanism and the calculation of the surface properties from the interference data. Another important factor in a WLI system is the interference objective that is used [72–74].

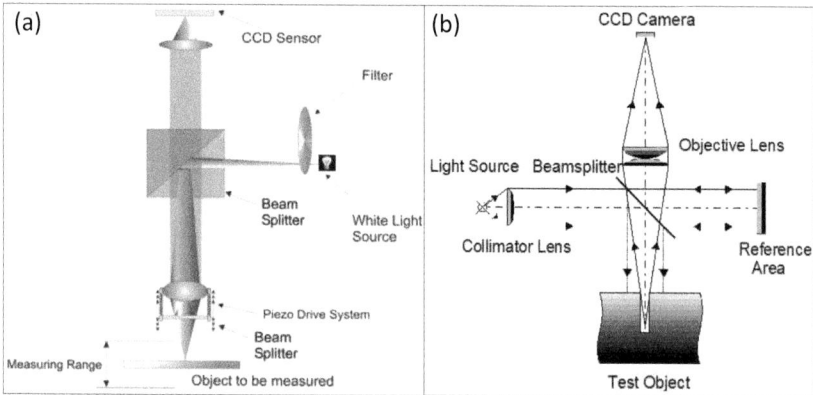

Figure 2.11 Principle and schematic diagram of a white light interferometer (WLI) system. (a) Principle of a WLI [71]; (b) schematic diagram of setting [72].

A low-magnification objective can be used to look at large areas but the resolution is controlled by the resolution of the detector. Higher resolution images need higher-magnification objectives but a smaller area has to be measured. The current lateral resolution limit for white light interferometry is about 0.5 μm because diffraction effects limit the maximum possible resolution. Another consideration when choosing an objective is the numerical aperture. The numerical aperture (NA) is related to the angle of the light that is collected by the objective. The higher NA then the greater the angle can be measured. Normally the higher-magnification objectives have a higher NA. Problems in white light interferometry can arise from the presence of thin films which can generate a second set of interference fringes. The two sets of fringes can cause errors in the analysis. In addition, materials with dissimilar optical properties can give an error in the measurement [72]. Accordingly, the Zygo white light interferometric profilometer offers fast, non-contact, high accuracy 3D metrology of surface features for a wide variety of samples. The software provides graphic images and high-resolution numerical analysis to characterize the surface structure of materials at magnifications up to 2000×. The maximum vertical range is 20 mm with a resolution of 0.1 nm [75].

2.1.2.2 Confocal optical microscopy

Confocal optical microscopy is one of the most widely used advanced techniques in surface metrology. It is called confocal because the microscope has two lenses having the same focus point, just as the name implies. The confocal microscope incorporates the ideas of point-by-point illumination of the specimen and rejection of out-of-focus light. The basic principle of the operation of the confocal microscope is discussed in references [76, 77] (see Fig. 2.12). The reflected beam reaches a diaphragm which transmits only focused light and to a photo detector. A vertical scanning system is moving the lens, which allows to analyze different height areas of specimen surface.

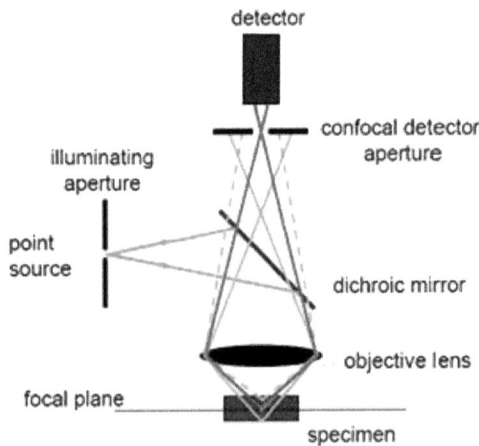

Figure 2.12 The principles of optical system of the scanning confocal microscope [78].

This ability to distinguish height improves significantly the contrast and the lateral resolution in comparison with the classic optical microscope. Scanning confocal microscopes take advantage of the differentiation of depth and generating of surface image and reception of reflected beam is done by the same optical system. Like in the scanning method, the optical system generates a spot on a surface, and a reflected light beam is recorded by a point detector.

2.1.2.3 Confocal white light microscopy

WLI and confocal microscopy seem good and particularly versatile. PSI is limited to smooth surfaces, whereas the vertical dynamic range of WLI and confocal microscopy extends from the nanometer level (noise) to a large range [67]. In connection with a suitably prepared light beam (after passing through another Nipkow disk with micro lenses), it allows a scan of the surface topography effectively. Construction with the confocal head connected to the traditional profilometers instead of the contact head has become more popular in recent years. The light illuminating the surface is usually white light. The beam incident on the surface is split into its spectrum by the passive optical system. Only one chosen frequency which depends on the height of every point is focused on it and gives a sharp image of this point on a photodetector [41]. The photo detector is a precise spectrophotometer which allows to identify that wavelength which gives the information about the height of the surface pattern of the measurement. The scheme of this construction is presented in Fig. 2.13.

Other scanning microscopes are applied to analyze surface of objects. Among these are two families of devices: scanning electron microscopes (SEM) and scanning probe microscopes (SPM). SEMs are used rarely in the investigation of roughness, whereas SPMs are becoming more and more popular. SEM is an optical method microscope technique that has been accepted in the nanoscale measurements. It allows obtaining very good vertical resolution. Serious problems appear only in the case of very rough surfaces and the necessity to use a larger vertical range. Microscope application can cause a little difficulty in the interpretation of results of measurement, especially when samples have inclusions or impurities on the surface with different characteristics or can have an influence on the response of instrument.

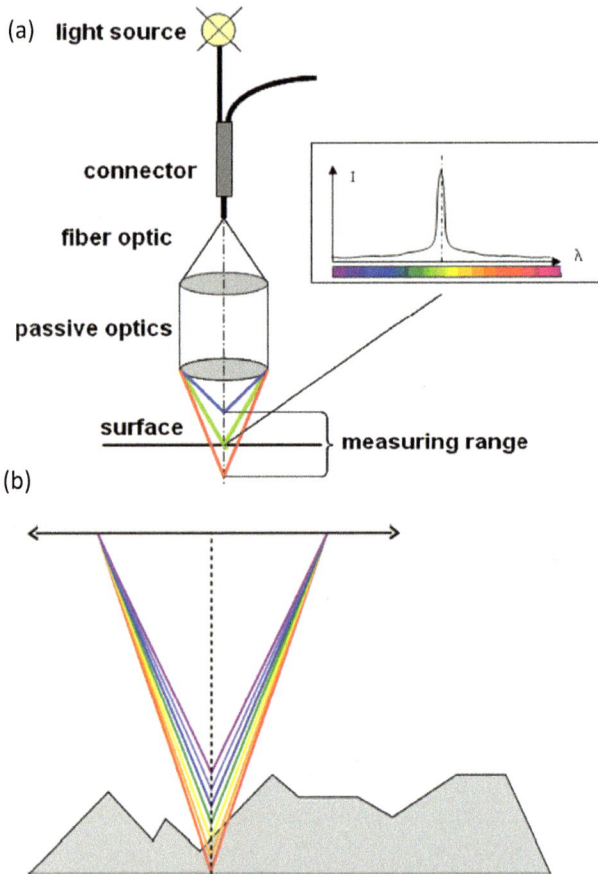

Figure 2.13 Schematic diagram of a confocal white light sensor for a profilometer. (a) schematic diagram [41, 79]; (b) measuring range principle [80].

2.1.2.4 Scanning electron microscopy

Scanning electron microscopy uses a focused beam of high-energy electrons to generate a variety of signals at the surface of specimens. The signals that derive from electron-sample interactions reveal information about the sample including external morphology such as chemical composition, crystalline structure and orientation of materials making up the sample.

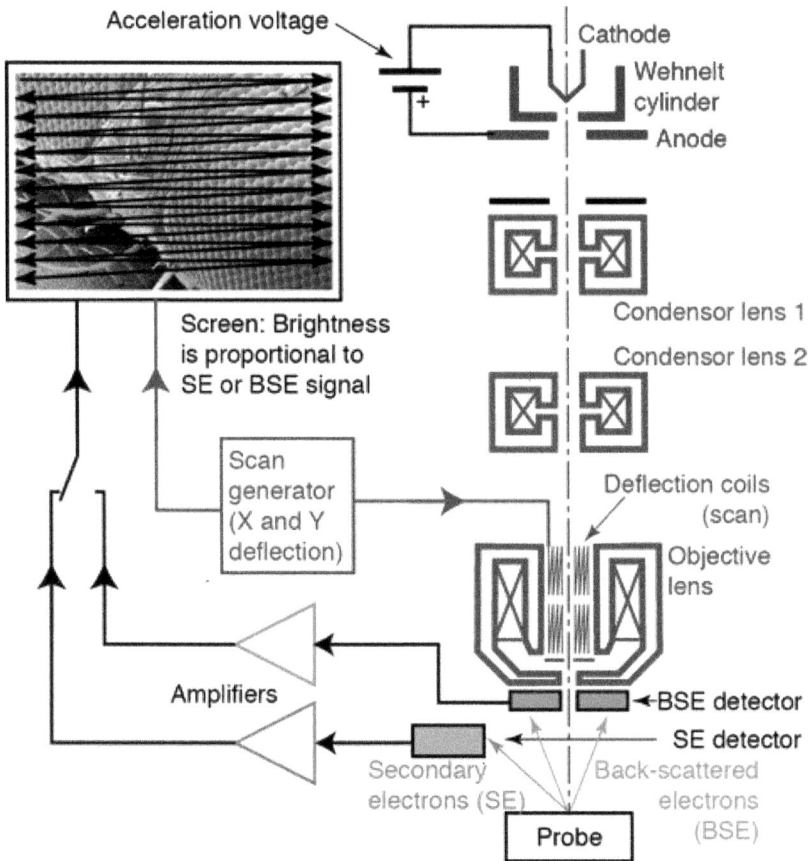

Figure 2.14 Schematic diagram of a SEM [81].

SEM is a type of electron microscopy that images a sample by scanning it with a beam of electrons in a raster scan pattern. The electrons interact with the atoms that make up the sample producing signals contain information about the sample's surface topography, composition, and other properties such as conductivity (Fig. 2.14). In most applications, data are collected over a selected area of the surface of the sample, and a 2D image is generated that displays spatial variations in these properties. The SEM creates an image by scanning the sample with a focused electron beam (a diameter of <5 nm is feasible) in a raster, the resolution typically 1, 2 μm [81]. Figure 2.15 illustrates the LEO 440 REM type of SEM [81].

Figure 2.15 The LEO 440 REM type of SEM [81].

2.1.2.5 Digital holography technique

Dennis Gabor invented holography as a method for recording and reconstructing the amplitude and phase of a wavefield in 1948. Digital holography (DH) technique is established as an important scientific tool for applications in imaging, microscopy, interferometry, and other optical disciplines [82, 83]. The DH setup for recording off-axis holograms is shown in Fig. 2.16,

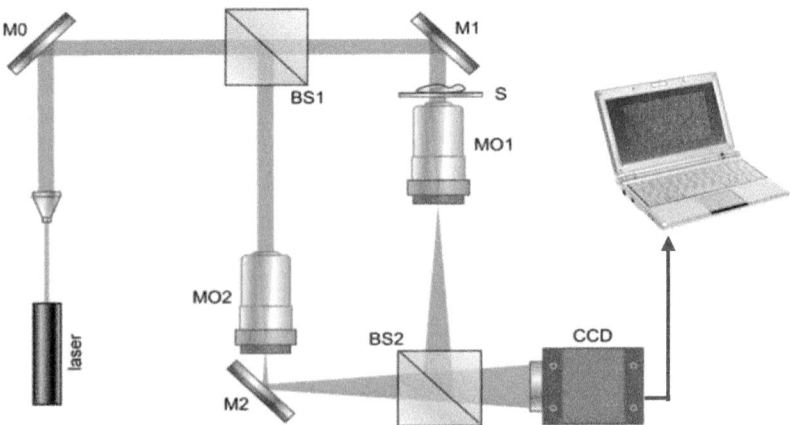

Figure 2.16 Schematic setup of digital holography technique [84].

where Ms are mirrors; BSs are beam splitters; MOs are microscope objectives and S is sample object. DH using based on collecting data via a CCD camera to investigate surface characteristics in some measurements applications such as engineering materials, biomedicine such as crack test. It consists of a Mach–Zehnder interferometer illuminated with a He–Ne laser whose wavelength is 633 nm [84].

2.2   Non-Optical Measurement Techniques

The SPM is a family of mechanical probe microscopes that scans the object in order to measure surface morphology in areal space with a resolution down to the atomic level [85]. An image of the surface obtained by mechanically moving the probe in a raster scan of the specimen, line by line, and recording the probe-surface interaction as a function of position. The two primary forms of SPM are scanning tunneling microscope (STM) and atomic force microscope (AFM). SPM was founded with the invention of the STM in 1981 by Binning and Rohrer for which they received the Nobel Prize for Physics in 1986 [86–88]. The STM is based on the concept of quantum tunneling, when the conductive tip is brought very near to the sample surface (below 1 nm) and applied bias electrons from the sample can tunnel through the vacuum between sample and tip. This tunnel electron flow is termed a tunneling current. The value of the intensity of this current decreases exponentially depending on the distance between tip and sample. Constant value of the tunneling current is maintained by feedback which controls distance between sample and tip. The STM registers changes in the value of the tunneling current with a constant distance or a change of the distance with the constant current. Besides STM and AFM, there are a number of different types of scanning probe microscopes, which can be classified in four types: optical, thermal, electric and force. Some of them are not separate types of microscope, but only modifications allowed using different physical forces [41, 89–92]. The STM and the AFM are the two most often used of the scanning non-optical probe microscopes. The fundamental application difference between these two types of microscopes is that STM can be used only for conductive materials but AFM can be used also for nonconducting materials [89]. The AFM microscope was the first constructed by Binning et al. [90] as a combination of a STM and a profilometer.

2.2.1 AFM Technique

Atomic force microscope is an advanced and important device in the family of SPMs as a non-optical measuring technique. The principle of AFM operation is based on surface scanning using an elastic cantilever with a sharp tip. The tip presses down to the surface with a small constant force. The tip has height from a small part of a micrometer up to 2 μm and tip radius from 2 to 60 nm. The interaction of the tip and the surface is monitored by the reflection of the laser beam from the top of a cantilever on the photodiode detector. Figure 2.17 shows the basic concept of AFM and STM.

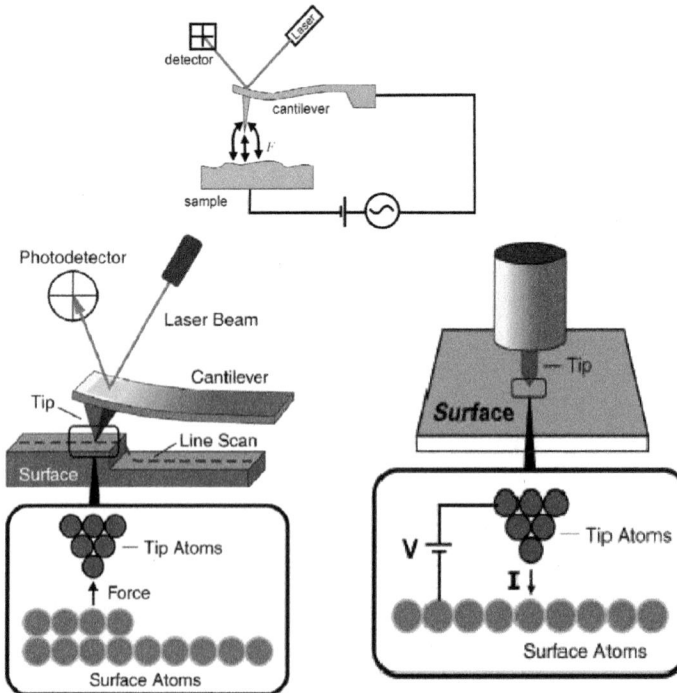

Figure 2.17 General principle of STM (left) and AFM (right) [88].

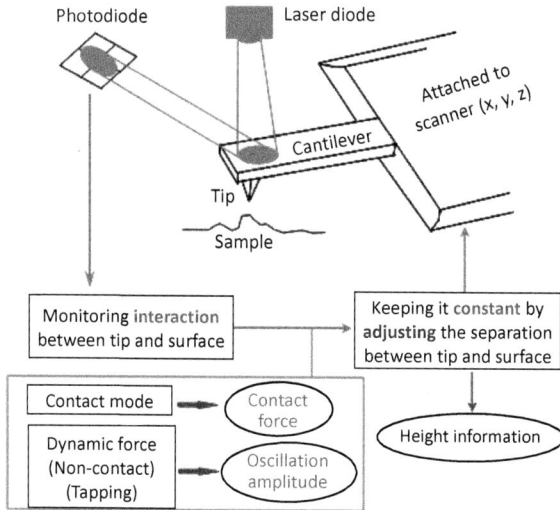

Figure 2.18 Operation of AFM system control loop [92].

There are some significant advantages of AFM as an imaging tool in surface metrology when compared with complementary techniques such as electron microscopy. In real time, AFM is currently able to achieve surface characterization of engineering nanomaterials and biomedicine more accurate than electron microscopes. AFM works at three different modes: non-contact mode, dynamic contact mode and tapping (resonant) mode. AFM measures the accurate forces acting between a fine probe tip and surface of an object sample. The tip attached to the free end of a cantilever and brought very close to a surface of sample. Attractive or repulsive forces resulting from interactions between the tip and the measured surface will cause a positive or negative bending of the cantilever. The bending is delegated by means of a laser beam, which is reflected from the backside of the cantilever. Figure 2.18 illustrates the block diagram in x, y, and z directions at different operation modes of AFM. While scanning the tip across the sample surface (x and y), the system adjusts the distance z (which is the measure of the height of the sample surface features) between the tip and the sample surface to maintain a constant contact force (contact mode) or oscillation amplitude (dynamic force mode). A 3D image is thus constructed by the lateral dimension the tip scans and the height the system measures. Figure 2.17 shows an example of AFM images obtained on a stamp in the 2D and the 3D.

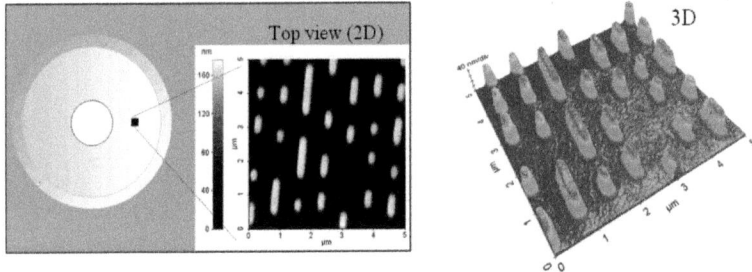

Figure 2.19 AFM image result in 2D and 3D [92].

2.2.2 3D-CT Technique

One of the new developed concepts in the recent years is the computer tomography (CT) metrology using X-rays. Computed tomography metrology is a technology to measure both internal and external geometries simultaneously in a great variety of parts. Therefore, the CT can be used not only as a simple inspection method but also as a measuring principle capable of providing accurate geometrical information. CT is in the process of revolutionizing quality engineering in industry [93–95]. The Metrotom CT-machine together with Calypso software is able to provide also internal and external 3D modeling of the measured part. The basic components of CT technique are illustrated in Fig. 2.20.

The CT machine consists of an X-ray source, a translational movable rotary table where the part to be scanned is placed on an X-ray detector and a processing unit (composed by six processors working together) to analyze and display the measurement results, Fig. 2.21. Currently two different types of CT systems exist depending on the beam and detector types [97, 98]. The 2D-CT systems use a fan beam and a line detector, whereas the 3D-CT systems utilize a cone beam and an area detector. In the case of the Metrotom CT-machine a 1024 × 1024 pixel (40 × 400 mm) in size area detector is used, so that 3D information can be measured with one revolution of the part. During the measurement processes the radiation which is not absorbed by the object is transmitted and recorded by the detector. For every angle rotated by the table, a new projection is obtained. Once a complete revolution takes place, the processing unit numerically reconstructs the measured object and provides a 3D graphical reconstruction. The primitive elements of the 3D data structure called voxels (volumetric pixels). Identification of surfaces using the CT-system makes it possible to determine coordinates of the measured part. Therefore, it is able to perform dimensional measurements like CMMs [94, 95].

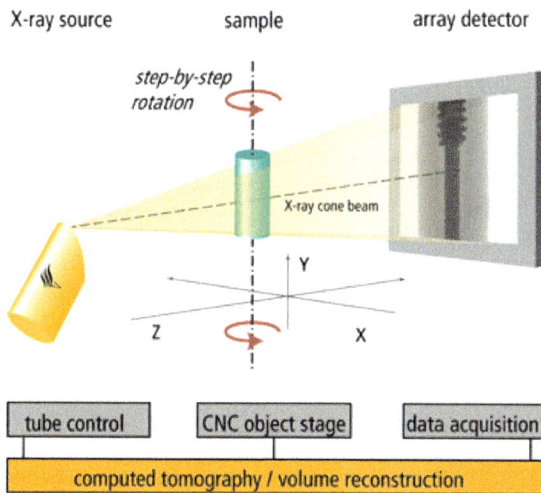

Figure 2.20 Principle operation of CT technique [96].

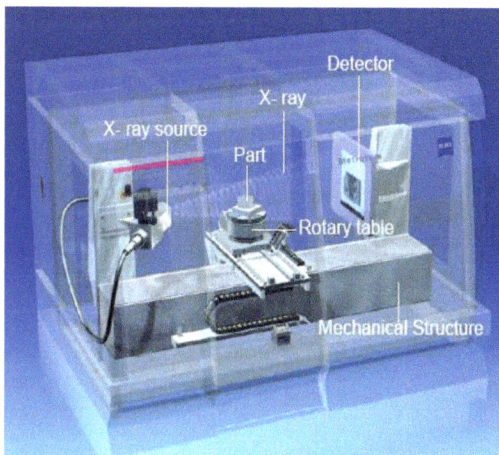

Figure 2.21 Components of Metrotom CT [94].

The CT technique provides an enormous amount of information and makes components transparent in the true sense inspection of machine spare parts. Several sources of error [94] can be associated to the detector properties, such as its lateral resolution, energy-dependent sensitivity (regulated by a gain ratio parameter), signal-to-noise ratio, dynamics, etc. The main errors are due to failed pixels that can be detected by an adequate calibration of the detector. The part itself is an important source of error from different points of view. Initially it must be adequately tightened to the rotary table. The roughness and cleanness of the part are important issues. The thermal expansion coefficient and the temperature during the process must be taken into account in the compensation part of the measurement software. The geometry and the thickness of the part have important influence on the quantity of radiation absorbed by the part, which can be different from the expected. The material composition will mainly determine the energy-dependent absorption (beamhardening), obtaining better results for materials absorbing less radiation (e.g., plastics). If the measured part is composed of a diversity of materials, unexpected and important effects can be obtained. The changes in the environmental conditions (vibrations, temperature changes) must be, of course, controlled and the system should be isolated from them as possible.

2.3   Overlapping, Limitations, Sampling, and Filtering of Existing Techniques

2.3.1   Overlapping

Over the past years, many metrological techniques of dimensions and surface measurements have matured and evolved as presented in Section 2.2. Consequently, the resolution overlapping between these instrumentation techniques is shown in Fig. 2.22.

It is obvious that there is a conjoint area between the measurement techniques, which spans a wide range and resolution. This indicates that there is generally a measurement solution for a wide variety of applications, including most that might be encountered in the engineering industries and applied technologies. The overlap between dimension and surface instrumentation techniques produces important benefits such as the following:

- provides competition between different technological techniques
- allows for a broader selection of instruments
- provides means for comparison

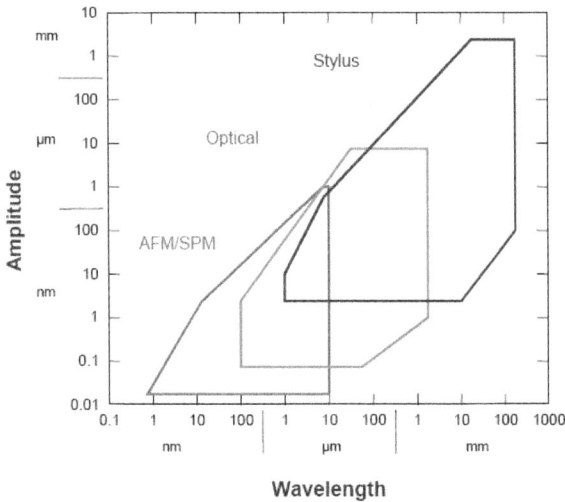

Figure 2.22 Steadman diagram showing the range and resolution of dimensional measurement techniques (mechanical stylus, interferomtric optical microscopes, and AFM/SPM) [99–101].

2.3.2 Limitations

The limitations of the existing techniques are diverse. Categories of the optical techniques are sometimes slightly artificial, not as in the case of the mechanical contact stylus techniques. Instruments which are the combination of several techniques have been created more frequently recently. It allows broadening the measurement range significantly while keeping very high resolution. An example of this solution is the interferometric microscope. General treatment of optical methods is very cautious, while the classic profilometric methods are used confidently. This situation is not incidental and results from many factors, e.g., application of the optical methods is sometimes questionable due to the fact that the estimation of the surface the whole mathematical models of the surface based on some assumption are used instead of using the surface itself. Moreover, results obtained from the optical methods depend sometimes on physical properties of the surface. In metals, for example, reflectivity is the significant parameter in contrast to some other materials where it is much lower, sometimes it is so low that a large amount of the incident light penetrates the material [41].

In the case of the layered surface, multiple reflections on different layers may occur. Diversity of the penetrations influences the optical length path and changes test results. The presence on the surface of elements which randomly disturb the light path—for example, small radius of curvature, micro cracking or micro holes—might be another reason for the abnormality. Further, the optical techniques cannot always be compared with stylus detection techniques, which sometimes make comparison of the test result impossible. Based on a number of comparative analyses, some practical limitations for stylus and optical techniques have been determined [41, 102, 103]. Heinrich Schwenke et al. discussed the differences between both of the technical limitations and the metrological characteristics of the methods [104]. Optical techniques like stylus methods require the isolation of devices from the external environment. Both thermal effects and vibrations have influence on the reliability of the result. Very careful cleaning of the sample surface is necessary from the point of view of industrial application. In summary, each device has its specific strategies in operation of measurement to avoid these limitations as much as possible.

2.3.3   Sampling and Filtering

Data collected from measurements is the starting point for many processes and may be used in different ways [41]. The data process of analogue-to-digital conversion, called digitization amounts to the representation of the continuous analogue signal by discrete data. In the frequency domain, the values of signal recorded at equal intervals in the plane of the surface. This process is called sampling. If the sampling interval is too small, the data are highly correlated and a large number of data points are required to represent surface topography. If it is too large, the resulting data are highly uncorrelated resulting in loss of surface spatial information (aliasing). In 3D stylus measurement, the sampling interval should be as large as possible, because measurement time (comparatively long) depends on it. An assessment of surface topography by parameters is useful when long wavelength components are removed from the measured surface data. The unwanted elements of the surface geometry are commonly referred to the waviness, due to imperfections in the manufacturing process. A necessary preliminary to numerical assessment of surface profiles is to extract the frequency components representative of the roughness and to eliminate those that would be irrelevant. A Gaussian filtering technique has been adopted for the filtration of surface topography. The Gaussian filtering technique solved the problems of phase distortion, but edge problems still exist in Gaussian filtration (marginal—running-in and running-out lengths, where roughness and waviness parameters cannot be calculated). The performance of the Gaussian filtering technique is affected by certain conditions, especially for surfaces having freak signals (outliers) such as grooves, scratches, and scores. Multi-process textures are an example of such surfaces. The problem is that control of such surface texture requires a complementary response from surface metrologists. Without adequate measurement technique the control and hence any attempt to maintain quality is lost. Therefore, there are many research efforts to study the effects of sampling and filtering of dimensional and surface metrology methods.

2.4 Surface Characterization

Many different characterization in the dimension and surface applications using advanced nanometrology techniques, whether 2D and 3D, in both micro- and nanometer scale can been presented. Many works present some new and promising approaches and may play an important role in industry and different applications using advanced metrology techniques. Thus, the macro engineering applications and surface porosity in the micro- and nanoscale have been presented.

2.4.1 Applications in the Mechanical Engineering

Surface metrology in mechanical engineering must be fast, accurate, robust, automated, and ideally integrated into the production line or product assessment, especially in materials science and automotive industry. With regard to the applications, CMMs offer a complete range of metrology solutions for many inspection tasks in different fields, in automotive industry and tribology. Some applications in CMM have been developed over the last years. The most familiar and earliest commercial application of surface texturing is that of automotive engine cylinder liner as a critical metallic element. Inspection of automotive mechanical parts designates the processes of measuring their dimensional and geometrical features has been presented in [105, 106] at NIS by the author of this work. The results of bore diameter and roundness deviations, as assessed, of the four cylinder liners of the test engine under investigation at 12 locations along the effective traveling stroke using CMM have been shown in [105] using a fraction of μm.

The topography of surfaces is commonly used to analyze surfaces after different operations and/or machining processes. The most familiar and earliest commercial application of surface texturing is that of automotive engine cylinder liner as a critical metallic element. At first, the recommended cylinder surface after honing was very smooth since it showed high wear resistance during running-in. Considerable progress in engine construction causes a great scuffing inclination of the smooth cylinder surface. Only rough surfaces have a higher load-carrying capacity. Running-in wear studies was conducted in ref. [107]. The compression ignition engine showed that the linear wear of the cylinder increased with increase in its initial surface roughness height. Rough surfaces wear rapidly without seizure during running-in period to promote quick performance, so an initial surface finish of the cylinder of 0.8 μm Ra (centerline average roughness) is recommended. A recently published work presents the surface texture as a real measured area of plateau roughness within the maintained Ra value, as shown in Fig. 2.23 [108].

The topography of engineering surfaces is commonly used to analyze surfaces before and after operation processes. The applications of texturing include; pistons, brake discs, bearings, mechanical face seals, gas seals, hard disk sliders, machine tools guideways and other elements. One can find recent publications presenting profitable effects of surface texturing on seizure resistance in refs. [109–111], transition between fluid and mixed lubrication in refs. [112–114], and wear resistance in refs. [115, 116]. The study of texturing usually requires multiscale analyses; therefore, different techniques have been used frequently [117, 118].

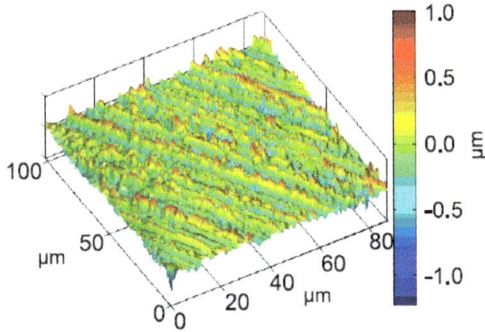

Figure 2.23 Plateau surface roughness using developed WLI technique [108].

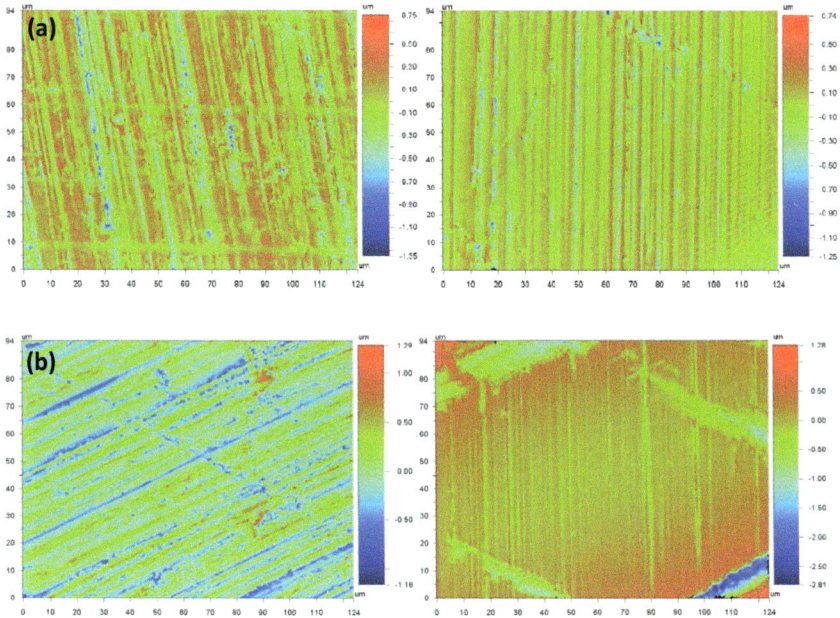

Figure 2.24 WLI images of [118]: (a) Piston ring (unworn section left, worn section right); (b) cylinder liner (left: unworn; right: worn).

Both the cylinder liner and the piston top compression ring have been measured with AFM and WLI [118]. Two different techniques are employed so that the effect of measuring technique on the results can be assessed. Figure 2.24 shows typical measurements for the four surfaces. The sliding motion in these images is in the vertical direction.

The piston ring is chromium coated and exhibits little wear apart from some minor scratching and smoothing in the direction of sliding. The cylinder liner is made of much softer cast iron and becomes markedly different after have been worn in the engine. The plateaux have been massively smoothed leaving the deep honing grooves clearly visible. Typical measurements for the four surfaces are given in Fig. 2.25. A similar trend of wear can be visually observing in the AFM measurements as for the WLI measurements.

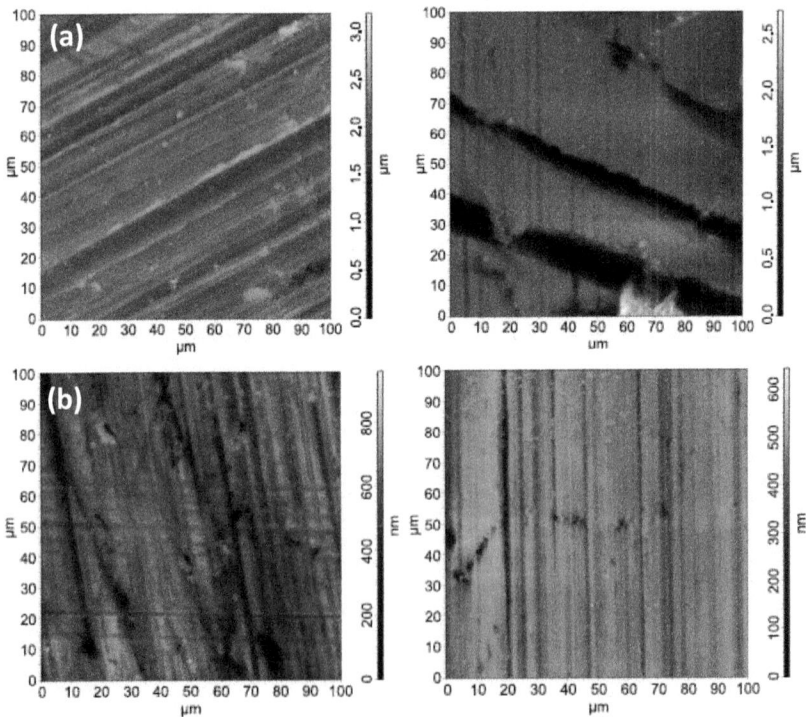

Figure 2.25 AFM images of [118]: (a) (engine piston ring (left: unworn; right: worn); (b) engine cylinder liner (left: unworn; right: worn).

2.4.2 Other Applications

Many other applications using surface metrology are presented. These fields of applications, such as road surface irregularity, architecture art tools, banknote paper, and biomaterial with tissue engineering using dimensional and surface metrology techniques, have been studied and analyzed. Thus, the inspection of concrete structures as a transportation infrastructure investment is a major part of roads management using surface metrology.

In surface metrology, macro structural application using SEM micrographs have been presented by the author of this work in [119, 120]. The objective of this work is to investigate the influence of doped and dispersed CNTs in polymer matrix on its intrinsic properties. Three different types of polymers—polyvinylchloride (PVC), polymethylmethacrylate (PMMA), and polystyrene (PS)—were subjected to this experimental investigation. CNTs/polymer matrix composites with a content ratio of CNTs up to 5% by weight were synthesized.

Figure 2.26 shows the SEM images of CNTs dispersion in three different types of polymer matrix composites. The apparently homogeneous dispersion and the strong interfacial bonding and cross-linking between CNTs and the PS polymer matrix are obvious. The functionalized knit-like interweave appearance indicates how CNTs play a role as a reinforcing agent in the PS matrix. It can be concluded that CNTs suit well the PS polymer matrix for synthesizing competent composite for different applications. The results of the thermal analysis of the CNTs/PS composites together with its FTIR spectrum discussed before, confirm this matching phenomenon of the CNTs and the PS polymer. It can be concluded that both PMMA and PS polymers showed much better matching ability and cross-linking ability with the interweaving CNTs than the PVC. Both PMMA and PS polymers may thus be nominated for further extension of the investigation to cover a wider range of CNTs content ratios in an endeavor toward searching for maximum specific performance properties of such nanocomposites.

Surface roughness of an architecture artist tool has been studied using the laser scanning microscope (LSM) technique [121, 122]. An advantage of LSM is that it uses a fine linear scale while obtaining the measurements. The linear scale is of 0.001 μm. The LSM has applications in various fields, mostly

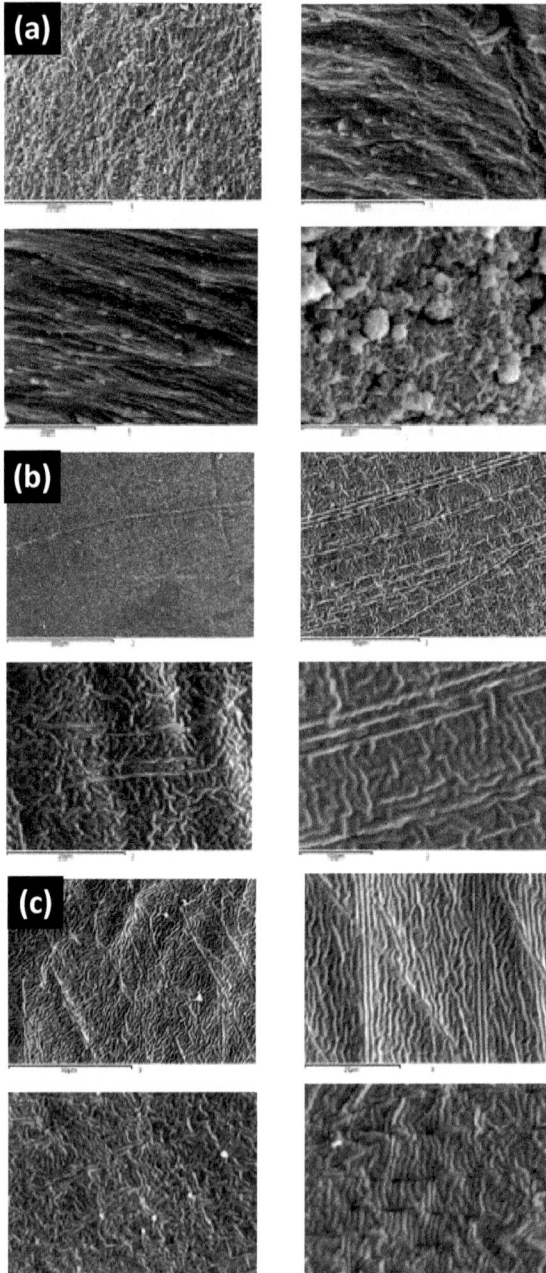

Figure 2.26 SEM images for nanocomposites [120]: (a) CNTs/PVC, (b) CNTs/PMMA, (c) CNTs/PS.

when we want to measure roughness or get information on color intensity. Figure 2.27 presents two different applications of LSM and give an idea about the result a user might receive upon using this microscope. The image on the left in Fig. 2.27 is a roughness measurement of a tool blade edge. Even though it can be actually seen from the results of the LSM measurements that there is a variance on the roughness. The same applies to the figure on the right, where they have measured the roughness of a silicon wafer. Again, the silicon wafer appears completely smooth.

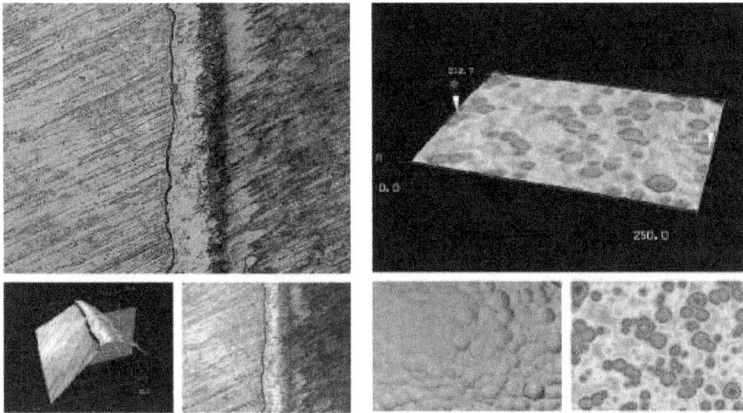

Figure 2.27 Various results obtained by using LSM [121, 122].

Surface topography analysis is often used in the paper and banknote industry. Because of the delicacy of the investigated surface, stylus measurements are replaced by other techniques mainly based on light scattering, although acoustic sensors [123] or AFM are also used. These methods allow for testing of paper for special applications, analysis of old prints and marks of physicochemical impurity or analysis connected with the deposition of a drop of ink from the printer. Roughness of the paper has influence on the deposition of the ink component, which causes non-uniformity of color printing [124]. On paper with higher roughness uniformity and quality of printing is worse. Figure 2.28 shows a surface of rough paper. Application of the topography parameters of the surface allows ascertaining that surfaces with similar values of some parameters may have totally different appearances and performances. Beside overprints on banknote surfaces, three-dimensional applied work is also plotted. Special marks make it difficult to make counterfeit banknotes and allow evaluating their authenticity. An image of a €50 banknote fragment is shown in Fig. 2.29.

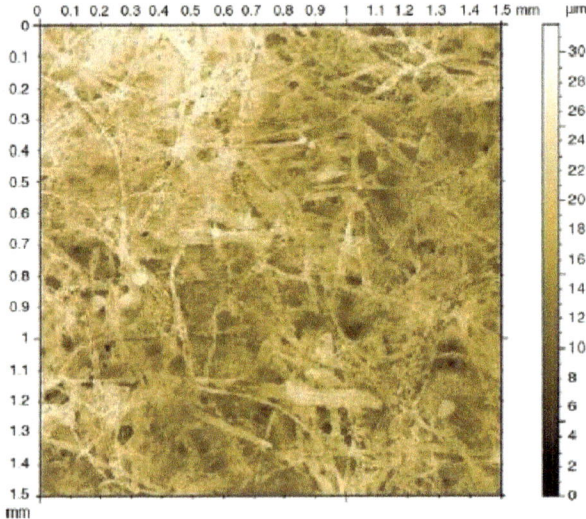

Figure 2.28   2D surface of rough paper [125].

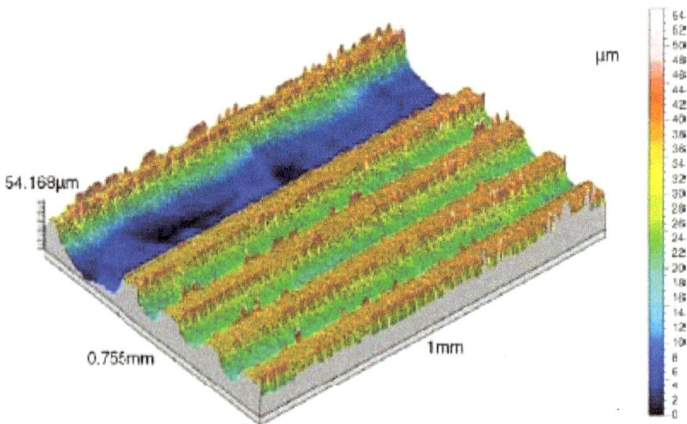

Figure 2.29   3D topography image of a banknote [40].

The next application is analysis of the surface of shot fired from the gun, which is used for gun identification [126]. In this branch, the NIST leads a program of model gun cartridge production, where it is fired from each gun directly after manufacturing and on which each gun leaves individual marks [127]. This cartridge is used for the identification of guns, because the marks of the gun barrel are impressed on its surface. The marks are measured as the roughness of each gun, which isas adequate as identification of fingerprints for people [128].

Through the development of products for the global automotive engineering, research applications of surface morphology have been used to support product development requirements. The use of high-velocity thermal spray as an alternative process for bearing manufacture to form plain bearings for automotive engine applications was developed. The microstructures of coated bearing cannot be obtained using conventional technique. SEM technique images a cross section through an AlSnCu powder coating layer using this method after heat treatment, with the lighter contrast phase, are shown in Fig. 2.30 [129]. The Al–Sn–Cu overlay is initially applied as a 100 mm coating over a bronze-based lining, then bored back to leave a thin layer of just a few microns, while also allowing the required dimensional tolerances to be achieved (see Fig. 2.31). The sprayed overlay has been successfully applied to both lead-containing and lead-free bronze linings [129].

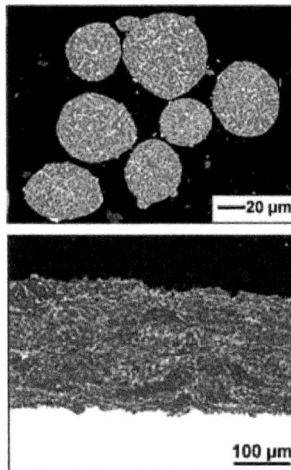

Figure 2.30 Photograph and SEM images (backscattered) of AlSnCu powder and coating heat treated at 300°C for 1 h. Both viewed in cross section.

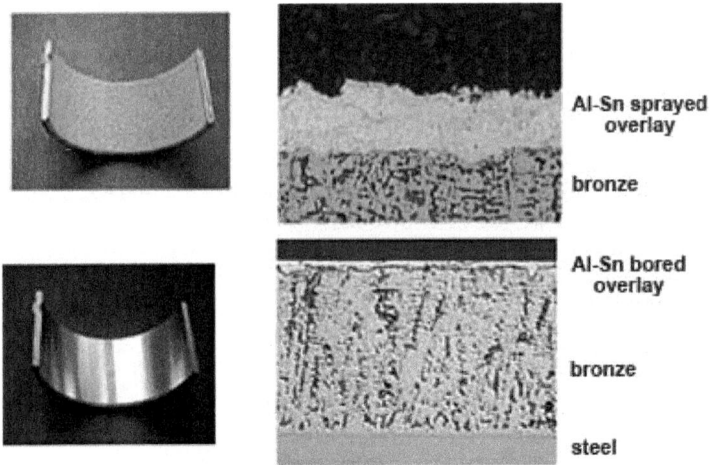

Figure 2.31 Spray deposited overlay on trimetal bearing.

2.5 Uncertainty

Uncertainty in measurement results of any metrological technique is considered as an important guide to the quality of the measurement procedure. One of the problems in evaluation of dimensional and surface measurements is that different factors affecting the total uncertainty cannot be separated. One can generally distinguish two types of uncertainty. The first one determines the systematic errors, which can be corrected theoretically. The second one determines the random errors. Recent research works involving the uncertainty resulting from measurement operations and their estimation in both contact and non-contact measurement methods have been carried out [130–134].

In CMM, Marsh et al. [130] compared the multi-step method with the Donaldson reversal method to test the ability of the methods to separate errors at the nanometer level. Abdelhak et al. [131] studied the uncertainty in measurement of CMM-based implementation of the multi-step method for the separation of machine and reference sphere errors on one side and triggering probe and probe tip errors on the other. Instead of a spindle rotating, the probe moves with the aid of the machine linear axes for each measurement. This method has been tested in the laboratory and the results were confirmed by measuring separately the probe errors and the machine errors. A good agreement between characteristics of the probe pre-travel shape was obtained between the proposed method and the independent measuring techniques. The work presented by Claudiu et al. [132] shows the uncertainty evaluation associated with the NPL areal new CMM. A sound mathematical model has been developed to describe the CMM geometry and functionality. Based on this model the uncertainty associated with the point coordinate measurement has been evaluated using a Monte Carlo technique, similar to the "virtual CMM" method. The Monte Carlo approach is based on using an input-output model and the probability distributions assigned to the input quantities in the model to evaluate the output quantity uncertainty. The main conclusion is that the NPL areal instrument is capable of measuring the relative position of a point on a surface with nanometer scale uncertainties. Intrinsic properties of the interferometers do not contribute significantly to the uncertainties associated with the point coordinate measured. In near future, to calibrate advanced accurate CMMs, further refinement of the mechanical parts of the laser tracking interferometer has to be introduced.

The uncertainty in the multi-sensorial measurement system consists of an optical light-section system and a shading system has been studied [133]. This work has been installed and assists in monitoring the process in real time better than before and leads to superior product quality and less rework at the end customer of the semi-finished products.

In spite of the information given by the previous studies of AFM tip-sample interactions, some problems are still unresolved in evaluating the uncertainty. An algorithm for the evaluation of the error and uncertainty contribution have been tested and developed with experimental results obtained from four step heights [134]. The application of the algorithm in this study is given for an aluminum rolling machine used in imprint lithography. Global uncertainty budget in AFM measurements, including calibration certificate, repeatability, orthogonally error, temperature drift and tip-sample angle parameters of the AFM measurements, has been estimated. Various methods for the evaluation of the uncertainty budget in measurement for dimensional measuring techniques can be found elsewhere [13, 26–38, 53, 132, 134].

2.6 Conclusion

In this chapter, a review of advanced measuring techniques in the dimensional and surface metrology and their applications from micro- to nanoscale has been presented in details. Recent progress and future trends under development in this area have been presented and discussed. The following are the major observations from the literature:

(1) The advanced nanomeasuring techniques are very important in the surface metrology for understanding the properties of the objects quality, design purposes, diagnostics, and high throughput screening. Both contact and non-contact metrology techniques are available methods for objects characterization.

(2) It is clear that the mechanical stylus instrument is viewed as the fundamental method of measurements and the mechanical surface analyses. However, now probe-scanning techniques are capable of characterizing the case of geometric surfaces in 3D.

(3) Optical nanomeasuring techniques have been reviewed recently because of their important role. For accurate topographic measurements with optical methods, there are a number of issues to be considered as slope limitations, smoothing, focus condition, stray light, surface detection algorithms, shadowing, and multiple scattering.

(4) Some important issues need urgent attention: acquisition of detected data in an economical and efficient way, filtering of noisy data, and extracting the statistical feature of the data.

(5) The hybrid opto-tactile probing system and nano ultra precision techniques need more research to reach high accuracy in the future.

(6) Multi integrated microscope as a combination between atomic force microscope and 3D optical profilometer needs advanced research to reach combine advantages in the future.

(7) The overlapping between dimensional measuring techniques needs new investigation in the context of presented advanced technology.

(8) With tremendous progress, the development of ISO standards becomes an essential requirement in the current time. This may require support and establishment of new technical committees.

If these issues are addressed, the engineering industrial application will become easy and fruitful task.

References

1. Salah H. R. Ali, Advanced Nanomeasuring Techniques for Surface Characterization, *ISRN Optics Journal*, vol. 2012, pp.1–23, Article ID 859353, 2012.

2. T. V. Vorburger, H.-G. Rhee, T. B. Renegar, J.-F. Song, and A. Zheng, Comparison of Optical and Stylus Methods for Measurement of Surface Texture, *The International Journal of Advanced Manufacturing Technology*, vol. 33, pp. 110–118, 2007.

3. ISO Committee Draft 25178-6, Geometrical Product Specification (GPS)-Surface Texture: Areal—Part 6: Classification of Methods for Measuring Surface Texture, 2007.

4. Ted Vorburger, ISO Standards for Surface Texture: Update, NIST, April 24, 2008. On line web site at: http://cstools.asme.org/csconnect/pdf/CommitteeFiles/29042.pdf.

5. J. A. Bosch, *Coordinate Measuring Machines and Systems*, Marcel Dekker, New York, 1995.

6. J. A. Bosch, *Coordinate Measuring Machines and Systems*, 1st ed., Marcel Dekker Inc., 1995.

7. J. Gou, *Theory and Algorithms for Coordinate Metrology*, PhD, Hong Kong University of Science and Technology, Jan. 1999.

8. Courtesy Xspect Solutions, Coordinate measuring machine (CMM) History. Online web site at: http://www.tusdec.org.pk/projects/gtdmc/solutions/tutorials/in/in-7.pdf.

9. Online at web page: http://www.caliperinc.com/new_page_8.htm.

10. J. Ni and F. Waldele, Coordinate Measuring Machine, Chapter 2, in *Coordinate Measuring Machine*, J. A. Bosch, ed., Marcel Dekker, New York, 1995.

11. S. H. R. Ali, The Influence of Fitting Algorithm and Scanning Speed on Roundness Error for 50 mm Standard Ring Measurement Using CMM, *The International Journal of Metrology & Measurement Systems*, vol. XV, no. 1, pp. 31–53, Jan. 2008.

12. International Standards: Geometrical Product Specifications (GPS)-Acceptance and Reverification Tests for Coordinate Measuring Machines (CMM)-Part 2: CMMs used for Measuring Size, ISO 10360-2, 2001.

13. S. D. Phillips, B. Borchardt, W. T. Estler, and J. Buttress, Estimation of Measurement Uncertainty of Small Circular Features Measured by CMMs. *Precision Engineering*, vol. 22, pp. 87–97, 1998.

14. S. H. R. Ali, Two Dimensional Model of CMM Probing System, *Journal of Automation, Mobile Robotics and Intelligent Systems*, vol. 4, no. 2, pp. 3–7, 2010.

15. S. H. R. Ali, Probing System Characteristics in Coordinate Metrology, *Journal of Measurement Science Review, Journal of the Institute of Measurement Science, Slovak Academy of Sciences*, vol. 10, no. 4, pp. 120–129, 2010.

16. G. Hermann, Geometric Error Correction in Coordinate Measurement, *Acta Polytechnica Hungarica*, vol. 4, no. 1, pp. 47–62, 2007.

17. J.-J. Park, K. Kwon, and N. Cho, Development of a Coordinate Measuring Machine (CMM) Touch Probe Using a Multi-Axis Force Sensor, *Measurement Science and Technology*, vol. 17, pp. 2380–2386, 2006.

18. A. Wozniak and M. Dobosz, Influence of Measured Objects Parameters on CMM Touch Trigger Probe Accuracy of Probing, *Precision Engineering*, vol. 29, no. 3, pp. 290–297, 2005.

19. A. Kasparaitis and A. Sukys, Dynamic Errors of CMM Probes, Diffusion and Defect Data, *Solid State Data, Part B*, ISSN 1012-0394, vol. 113, pp. 477–482, 2006.

20. Y. Wu, S. Liu, and G. Zhang, Improvement of Coordinate Measuring Machine Probing Accessibility, *Precision Engineering*, vol. 28, pp. 89–94, 2004.

21. J.-A. Yague, J.-A. Albajez, J. Velazquez, and J.-J. Aguilar, *A New Out-of-Machine Calibration Technique for Passive Contact Analog Probes*, Elsevier Ltd., Measurement, vol. 42, pp. 346–357, 2009.

22. A. Wozniak, J. R. R. Mayer, and M. Balazinski, Stylus Tip Envelop Method: Corrected Measured Point Determination in High Definition coordinate metrology, *The International Journal of Advanced Manufacturing Technology*, vol. 42, pp. 505–514, 2009.

23. M. Antonetti, M. Barloscio, A. Rossi, and M. Lanzetta, Fast Genetic Algorithm for Roundness Evaluation by the Minimum Zone Tolerance (MZT) Method, 2011. On line web site at: http://eprints.adm.unipi.it/1223/2/TEXT26_as_submitted.pdf.

24. Zeiss Calypso Navigator, CMM Operation Instructions and Training Manual, Revision 4.0, Germany, 2004.

25. A. Garces, D. Huser-Teuchert, T. Pfeifer, P. Scharsich, F. Torres-Leza, and E. Trapet, Performance Test Procedures for Optical Coordinate Measuring Probes, Report on Contract no. 3319/1/0/159/89/8-BCR-D(30), European Communities, Brussels, 1993.

26. R. Donker, I. Widdershoven, and H. Spaan, Realization of Isara 400: A Large Measurement Volume Ultra-Precision CMM, *Asian Symposium for Precision Engineering and Nanotechnology*, 2009.

27. Web site on: http://www.meopta.com/en/o-inspect-1404042583.html.

28. Z. G. Fei, J. J. Guo, X. J. Ma, D. Z. Gao, A novel type compact five-coordinate measuring machine with laser and CCD compound probe, accepted for publishing in Measurement, 2011.

29. J. Sładek, P. M. Błaszczyk, M. Kupiec, and R. Sitnik, Review: The Hybrid Contact-Optical Coordinate Measuring System, *Measurement*, vol. 44, pp. 503–510, 2011.

30. L. Xiuming, and S. Zhaoyao, Application of Convex Hull in the Assessment of Roundness Error, *International Journal of Machine Tools and Manufacture*, vol. 48, no. 6, pp. 711–714, 2008.

31. T.-H. Sun, Applying Particle Swarm Optimization Algorithm to Roundness Measurement, *Expert Systems with Applications*, vol. 36, no. 2, Part 2, pp. 3428–3438, March 2009.

32. D. Bogomolov, V. Poroshin, A. Kostyuk, and V. Radygin, High Precision PC Based Measurement Systems for Complex Analysis of roundness and waviness, *Advanced Engineering*, ISSN 1846-5900, 2010.

33. Online:http://www.rosebank-eng.com.au/news~latest_news~TR365.html.

34. Talyround Measuring Technique, TR-73. Online web site at: http://www.npl.co.uk/engineering-measurements/dimensional/dimensional-measurements/products-and-services/high-accuracy-roundness-measurement.

35. Online web site at: http://taylor-hobson.virtualsite.co.uk/ultraroundness.htm.

36. W. Gao, et al., On-Machine Roundness Measurement of Cylindrical Workpieces by the Combined Three-Point Method, *Precision Engineering*, vol. 21, no. 4, pp. 147–156, 1997.

37. S. Mekid and K. Vacharanukul, In-Process Out-of-Roundness Measurement Probe for Turned Workpieces, *Measurement*, vol. 44, pp. 762–766, 2011.

38. R. Verma and J. Raja, Characterization of Engineered Surfaces, *Journal of Physics: Conference Series*, vol. 13, pp. 5–8, 2005.

39. E. J. Abbott and F. A. Firestone, Specifying Surface Quality, *Mechanical Engineering*, vol. 55, pp. 569–572, 1933.

40. T. G. Mathia, P. Pawlus, and M. Wieczorowski, Recent Trends in Surface Metrology, *Wear*, vol. 271, pp. 494–508, 2011.

41. 8000-T Nanoscan, Hommel-Etamic Jenoptik, Schwenningen, 2008. On line web site at: http://www.jenoptik.com/Internet_EN_HOMMEL-ETAMIC_T8000_wavemove.

42. E. Peiner, Compliant Tactile Sensors for High-Aspect-Ratio Form Metrology, Chpter 21, in: *Sensors, Focus on Tactile, Force and Stress Sensors*, J. G. Rocha and S. Lanceros-Mendez, ISBN 978-953-7619-31-2, pp. 444, I-Tech, Vienna, Austria, Dec. 2008.

43. E. Peiner, M. Balke, and L. Doering, Form Measurement Inside Fuel Injector Nozzle Spray Holes, *Microelectronic Engineering*, vol. 86, pp. 984–986, 2009.

44. M. Wieczorowski, Spiral Sampling as a Fast Way of Data Acquisition in Surface Topography, *International Journal of Machine Tools and Manufacture*, vol. 41, pp. 2017–2022, 2001.

45. J. Song, The Effect of Gaussian Filter Long Wavelength Cutoff_c in Topography Measurements and Comparisons, in: *Proc XXI ASPE AM*, pp. 387–390, 2006.

46. J. M. Bennett, Overview: Sensitive Techniques for Surface Measurement and Characterization, *Proceedings of SPIE*, 1573, pp. 152–158, 1991.

47. J. M. Bennett, Recent Developments in Surface Roughness Characterization, *Measurement Science and Technology*, vol. 3, pp. 1119–1127, 1992.

48. H. J. Tiziani, Optical Methods for Precision Measurements, *Optical and Quantum Electronics*, vol. 21, pp. 256–282, 1989.

49. K. Leonhardt, et al., Optical Methods of Measuring Rough Surfaces, *Proceedings of SPIE*, 1009, pp. 22–29, 1988.

50. C. J. Tay, S. Wang, C. Quan, and H. M. Shang, In-Situ Surface Roughness Measurement Using a Laser Scattering Method, *Optics Communications*, vol. 218, no. 1–3, pp. 1–10, 2003.

51. S. Wang, Y. H. Tian, C. J. Tay, and C. Quan, Development of a Laser Scattering Based Probe for on-Line Measurement of Surface Roughness, *Applied Optics*, vol. 42, no. 7, pp. 1318–1324, 2003.

52. R. Ohisson, A Topographic Study of Functional Surfaces, PhD Thesis, Chalmers University of Technology, 1996.

53. UKAS, Measurement uncertainty definitions. Online at web page: http://www.ukas.com/technical-information/publications-and-tech-articles/technical/technical-uncertain.asp.

54. A. Harasaki, et al., Improved Vertical Scanning Interferometry, *Applied Optics*, vol. 39, pp. 2107–2115, 2000.

55. B. Bhushan, J. C. Wyant, and C. L. Koliopoulis, Measurement of Surface Topography of Magnetic Tapes by Mirau Interferometry, *Applied Optics*, vol. 24, pp. 1489–1497, 1985.

56. J. E. Greivenkamp, and J. H. Bruning, In: D. Malacara, ed., *Optical Shop Testing*. Wiley, New York, pp. 501–598, 1992.

57. Y. N. Denisyuk, On the Reproduction of the Optical Properties of an Object by the Wave Field of Its Scattered Radiation, Pt. I, *Optics and Spectroscopy (USSR)*, vol. 15, p. 279, 1963.

58. Y. N. Denisyuk, On the Reproduction of the Optical Properties of an Object by the Wave Field of Its Scattered Radiation, Pt. II, *Opt. Spectroscopy (USSR)*, vol. 18, p. 152, 1965.

59. B. J. Chang, R. C. Alferness, and E. N. Leith, Space-Invariant Achromatic Grating Interferometers: Theory (TE), *Applied Optics*, vol. 14, p. 1592, 1975.

60. E. N. Leith and G. J. Swanson, Achromatic Interferometers for White Light Optical Processing and Holography, *Applied Optics*, vol. 19, p. 638, 1980.

61. Y.-S. Cheng and E. N. Leith, Successive Fourier Transformation with an Achromatic Interferometer, *Applied Optics*, vol. 23, p. 4029, 1984.

62. E. N. Leith and R. R. Hershey, Transfer Functions and Spatial Filtering in Grating Interferometers, *Applied Optics*, vol. 24, p. 237, 1985.

63. J. C. Wyant, et al., An Optical Profilometer for Surface Characterization of Magnetic Media, *ASLE Transactions*, vol. 27, pp. 101–113, 1984.

64. P. F. Forman, The ZYGO Interferometer System, *Proceedings of SPIE*, vol. 129, pp. 41–48, 1979.

65. L. Deck and P. deGroot, High-Speed Noncontact Profiler Based on Scanning White-Light Interferometer, *Applied Optics*, vol. 33, pp. 7334–7388, 1994.

66. J. Schmit, A. Olszak, High-Precision Shape Measurement by White-Light Interferometry with Real-Time Scanner Correction, *Applied Optics*, vol. 41, pp. 5943–5950, 2002.

67. T. V. Vorburger, H.-G. Rhee, T. B. Renegar, J.-F. Song, and A. Zheng, Comparison of Optical and Stylus Methods for Measurement of Surface Texture, *The International Journal of Advanced Manufacturing Technology*, vol. 33, pp. 110–118, 2007.

68. J. C. Wyant, White Light Extended Source Shearing Interferometer, *Applied Optics*, vol. 13, pp. 200–203, 1974.

69. Y.-Y. Cheng and J. C. Wyant, Two-Wavelength Phase Shifting Interferometry, *Applied Optics*, vol. 23(24), pp. 4539–4543, 1984.

70. Y.-Y. Cheng and J. C. Wyant, Multiple-Wavelength Phase-Shifting Interferometry, *Applied Optics*, vol. 24, no. 6, pp. 804–807, 1985.

71. Optical Topography Measurement and Material Characterisation of Alu Cylinder Faces. Online web site at: http://www.breitmeier.com/component/content/article/119-wissen#fig6.

72. Mike Conroy and Joe Armstrong, A Comparison of Surface Metrology Techniques, 7th International Symposium on Measurement Technology and Intelligent Instruments, *Journal of Physics: Conference Series*, vol. 13, pp. 458–465, 2005.

73. S. Ma, C. Quan, R. Zhu, C. J. Tay, and L. Chen, Surface Profile Measurement in White-Light Scanning Interferometry Using a Three-Chip Color CCD, *Applied Optics*, vol. 50, no. 15, pp. 2246–2254, 2011.

74. S. Ma, C. Quan, R. Zhu, C. J. Tay, L. Chen, and Z. Gao, Micro-Profile Measurement Based on Windowed Fourier Transform in White-Light Scanning Interferometry, *Optics Communications*, vol. 284, no. 10, pp. 2488–2493, 2011.

75. Zygo-3D, web sit: http://www.ims.uconn.edu/facilities/surface_analysis.html.

76. M. Minsky, Microscopy Apparatus, U.S. Patent no. 3013467, 19, December 1961.

77. M. Minsky, Memoir on Inventing the Confocal Scanning Microscope, *Scanning*, vol. 10, no. 4, pp. 128–138, 1988.

78. On line web site at: http://www.gum2012.fionastoreydesign.co.uk/confocal_microscopy.html.

79. Confocal white light sensor, Hommel-Etamic Jenoptik, Schwenningen, 2007.

80. Craig Leising, Accurate Micropart Topography Using 3-D Metrology, 12/02/2010. Available at: http://www.qualitydigest.com/inside/

metrology-article/accurate-micropart-topography-using-3-d-metrology.html.

81. Petrology and Geochemistry Laboratory, Heidelberg University, Publishing Information. Website at: http://www.rzuser.uni-heidelberg.de/~hb6/labor/rem/index_en.html.

82. Ulf Schnars and Werner P. O. Jüptner, Review Article: Digital Recording and Numerical Reconstruction of Holograms, *Measurement Science and Technology*, vol. 13, pp. R85–R101, 2002.

83. N. F. A. Maaboud, et al., Digital Holography in Flatness and Crack Investigation, *Metrology and Measurement Systems*, vol. XVII, no. 4, pp. 583–588, 2010.

84. X. Yu, et al., Quantitative Imaging and Measurement of cell–substrate Surface Deformation by Digital Holography, *Journal of Modern Optics*, pp. 1–8, iF, 2012.

85. A. Vilalta-Clemente and K. Gloystein, Principles of Atomic Force Microscopy (AFM), Based on the Lecture of Prof. Nikos, Frangis Aristotle University, Thessaloniki, Greece, Physics of Advanced Materials Winter School, 2008. Online at web page: http://www.mansic.eu/documents/PAM1/Frangis.pdf.

86. Ernst Ruska, Ernst Ruska Autobiography, Nobel foundation. Online at web site: http://nobelprize.org/nobel_prizes/physics/laureates/1986/ruska-autobio.html Retrieved 2010-01-31, 1986.

87. G. Binnig, C. F. Quate, and Ch. Gerber, Atomic Force Microscope, *Physical Review Letters*, vol. 56, no. 9, pp. 930–933, 3 March 1986.

88. The principles of Atomic Force Microscopy (AFM). Online at web page: https://www.uclan.ac.uk/schools/computing_engineering_physical/jost/files/AFM.pdf.

89. Difference between AFM and STM, 2011. Online web site at: http://www.differencebetween.net/technology/difference-between-afm-and-stm/.

90. G. Binning, et al., Atomic Force Microscope, *Physical Review Letters*, vol. 56, pp. 930–933, 1986.

91. G. Binning, et al., Surface Studies by Scanning Tunneling Microscopy, *Physical Review Letters*, vol. 49, pp. 57–61, 1982.

92. H.-Y. Nie, Scanning Probe Techniques. Online at web page: http://publish.uwo.ca/~hnie/spmman.html#c-afm.

93. M. Simon, I. Tiseanu, and C. Sauerwein, Advanced Computed Tomography System for the Inspection of Large Aluminium Car Bodies, ECNDT, 2006.

94. V. Andreu, B. Georgi, H. Lettenbauer, and J. A. Yague, Analysis of the Error Sources of a Computer Tomography Machine, pp. 1–10, 2009. Available at: http://www.iberprecis.es/docs/Paper_CT.pdf.

95. S. Kasperl, J. Hiller, and M. Krumm, Computed Tomography Metrology in Industrial Research & Development, International Symposium on NDT in Aerospace, Germany, pp. 1–8, December 2008.

96. Availableat:http://www.phoenix-xray.com/en/company/technology/principles_of_operation/principle_060.html.

97. M. Bartscher, U. Neuschaefer-Rube, and F. Waldele, Computed Tomography-A highly Potential Tool for Industrial Quality Control and Production Near Measurements, *VDI-Berichte*, vol. 1860, pp. 477–482, 2004.

98. M. Bartscher, U. Hilpert, J. Goebbels, and G. Weidemann, Enhancement and Proof of Accuracy of Industrial Computed Tomography (CT) Measurements, *Annals of the CIRP*, vol. 56, pp. 495–498, 2007.

99. L. Brown and B. Liam, Surface Metrology for the Automotive Industry. In: Inaugural Automotive Researchers' Conference, University of Huddersfield, 8–9 Jan. 2008. On line at: http://eprints.hud.ac.uk/4059/1/08AARC2008.pdf.

100. Christopher, Basic Elements of the Design and Selection of Surface Metrology Systems, 2006. On line web site at: http://www.me.wpi.edu/Images/CMS/SML/Christopher_Brown_Surf_Met_Seminar_WPI_tutorial_06.pdf.

101. C. A. Brown, The Geometric Nature of Surface Roughness and Microwave Transmission Losses. Online at: http://www.imapsne.net/archive/docs/Roughness-%20Brown.pdf.

102. M. Stedman, Limits of Surface Measurement by Optical Probes, *Proceedings SPIE*, vol. 1009, pp. 62–67, 1988.

103. M. Stedman and K. Lindsey, Limits of Surface Measurement by Stylus Instruments, *Proceedings SPIE*, 1009, pp. 56–61, 1988.

104. H. Schwenke, U. Neuschaefer-Rube, T. Pfeifer, and H. Kunzmann, Optical Methods for Dimensional Metrology in Production Engineering, *CIRP Annals–Manufacturing Technology*, vol. 51, no. 2, pp. 685–699, 2002.

105. S. H. R. Ali, M. K. Bedewy, and S. Z. Zahwi, Dimensional Inspection of Overhauled Automotive Water-Cooled Diesel Engines, *9th International Conference on Production Engineering, Design, and Control PEDAC'09*, Alexandria, Egypt, February 10–12, 2009.

106. S. H. R. Ali, H. H. Mohamed, and M. K. Bedewy, Identifying Cylinder Liner Wear Using Precise Coordinate Measurements, *International*

Journal of Precision Engineering and Manufacturing, vol. 10, no. 5, pp. 19–25, Dec. 2009.

107. A. V. Sreenath, N. Raman, Running-in Wear of a Compression Ignition Engine: Factors Influencing the Conformance between Cylinder Liner and Piston Rings, *Wear*, vol. 38, pp. 271–289, 1976.

108. A. Spencer, A. Almqvist, and R. Larsson, A Semi-Deterministic Texture-Roughness Model of the Piston Ring-Cylinder Liner Contact, *Proceedings of the Institution of Mechanical Engineers, Part J: Journal of Engineering Tribology*—Leeds-Lyon Special Issue, Received 21st July 2010.

109. L. Rapoport, A. Moskovith, V. Perfilyev, I. Lapsker, G. Halperin, Y. Itovich, and I. Etsion, Friction and Wear of MoS2 Films on Laser Textured Steel Surfaces, *Proceedings of NORDTRIB Symposium*, Tampere, Finland, 2008.

110. W. Koszela, L. Galda, A. Dzierwa, and P. Pawlus, The Effect of Surface Texturing on Seizure Resistance of Steel-Bronze Assembly, 36th Leeds-Lyon Symposium on Tribology, Lyon, France, 2009.

111. L. Galda, A. Dzierwa, J. Sep, and P. Pawlus, The Effect of Oil Pockets Shape and Distribution on Seizure Resistance in Lubricated Sliding, *Tribology Letters*, vol. 37, no. 2, pp. 201–211, 2010.

112. E. Gualtieri, A. Borghi, L. Calabri, N. Pugno, and S. Valeri, Increasing Nanohardness and Reducing Friction of Nitride Steel by Laser Surface Texturing, *Tribology International*, vol. 42, pp. 699–705, 2009.

113. L. Galda, P. Pawlus, and J. Sep, Dimples Shape and Distribution Effect on Characteristics of Stribeck Curve, *Tribology International*, vol. 42, pp. 1505–1512, 2009.

114. L. M. Vilhena, M. Sedlacek, B. Podgornik, J. Vizintin, A. Babnik, and J. Mozina, Surface Texturing by Pulsed Nd: YAG Laser, *Tribology International*, vol. 42, pp. 1496–1504, 2009.

115. W. Koszela, P. Pawlus, and L. Galda, The Effect of Oil Pockets Size and Distribution on Wear in Lubricated Sliding, *Wear*, vol. 263, pp. 1585–1592, 2007.

116. M. Mosleh and B. A. Khemet, A Surface Texturing Approach for Cleaner Disc Brake, *Tribology and Lubrication Technology*, vol. 62, no. 12, pp. 32–37, Dec. 2006.

117. F.-P. Ninove, C. Rapiejko, and T. G. Mathia, 3D Morphological Behaviour of Heterogeneous Model Material in Finishing Abrasive Process-Case of Nodular Cast Iron, in: J. Sladek and W. Jakubiec, eds., *Advances in Coordinate Metrology*, University of Bielsko-Biala, pp. 456–462, ISBN 978-83-62292-56-1, 2010.

118. A. Spencer, I. Dobryden, N. Almqvist, A. Almqvist, and R. Larsson, Surface Characterization with Functional Parameters, *STLE: Tribology Transactions*, 2010.

119. S. H. R. Ali, M. A. Etman, B. S. Azzam, R M., Rashad, and M. K. Bedewy, Advanced Nanometrology Techniques for Carbon Nanotubes Characterization, *International Journal of Metrology & Measurement Systems, Warsaw, Poland*, vol. XV, no. 4, pp. 551–561, October 2008.

120. S. H. R. Ali, M. K. Bedewy, M. A. Etman, H. A. Khalil, and B. S. Azzam, Morphology and Properties of Polymer Matrix Nanocomposites, *International Journal of Metrology and Quality Systems*, vol. 1, no. 1, pp. 33–39, France, 2010.

121. Keyence, Keyence World Class Products for Sensing and Automation, 2008.

122. C. Altin and I. Ozil, Applications of Surface Metrology to Issues in Art, Project Report, Worcester Polytechnic Institute, 2008. Online web site at: http://www.wpi.edu/pubs/e-project/available /e-project-042408-121857/unrestricted/iqp_ceren_ipek.pdf.

123. J. Stor-Pellinen and M. Luukkala, Paper Roughness Measurement Using Airborne Ultrasound, *Sensors and Actuators A*, vol. 49, pp. 37–40, 1995.

124. J. S. Preston, et al., Investigation into the Distribution of Ink Components on Printed Coated Paper, Optical and Roughness Considerations, *Colloids and Surfaces A*, vol. 205, pp. 183–198, 2002.

125. M. Zecchino, Characterizing Surface Quality: Why Average Roughness Is Not Enough, *Advanced Materials and Processes*, vol. 161, pp. 25–28, 2003.

126. N. Senin, et al., Three-Dimensional Surface Topography Acquisition and Analysis for Firearm Identification, *Journal of Forensic Sciences*, vol. 51, pp. 282–295, 2006.

127. E. P. Whitenton, et al., Manufacturing and Quality Control of the NIST Reference Material 8240 Standard Bullet, in: *Proceedings of XVIII ASPE AM*, pp. 99–102, 2003.

128. J. Song, et al., Standard Bullets and Casings Project, SRM 2460/2461, *Journal of research of the National Institute of Standards and Technology*, vol. 109, pp. 533–542, 2004.

129. A. J. Sturgeon, C. Perrin, and D. G. McCartney, Development of Thermal Sprayed Plain Bearings for Automotive Engine Applications, *Tribology 2006: Surface Engineering & Tribology for Future Engine and Drivelines*, IMechE, London, 12–14 July 2006.

130. E. Marsh, J. Couey, and R. Vallance, Nanometer-Level Comparison of Three Spindle Error Motion Separation Techniques, *Journal of Manufacturing Science and Engineering*, vol. 128, pp. 180–187, 2006.

131. A. Nafi, J. R. R. Mayer, and A. Wozniak, Novel CMM-Based Implementation of the Multi-Step Method for the Separation of Machine and Probe Errors, *Precision Engineering*, vol. 35, pp. 318–328, 2011.

132. C. L. Giusca, R. K. Leach, and A. B. Forbes, A Virtual Machine-Based Uncertainty Evaluation for a Traceable Areal Surface Texture Measuring Instrument, *Measurement*, vol. 44, pp. 988–993, 2011.

133. J. Bernstein and A. Weckenmann, User Interface for Optical Multi-Sensorial Measurements at Extruded Profiles, *Measurement*, vol. 44, pp. 202–210, 2001.

134. H. Gonzalez-Jorge, M. A. Fernandez-López, J. L. Valencia, and S. Torres, Uncertainty Contribution of Tip-Sample Angle to AFM Lateral Measurements, *Precision Engineering*, vol. 35, pp. 164–172, 2011.

PART 3

PERFORMANCE OF CMM METROLOGY TECHNIQUE

Chapter 3

Characterization of Touch Probing System in CMM Machine

Coordinate measuring machine is widely utilized as a precise dimensional measuring tool in the modern manufacturing processes, especially in the mass production, as in automotive industries. CMM probe is one of the most important systems of dimensional measuring instruments and responsible for the coordinate measurement accuracy [1–5]. Fast response and accurate detection of probe that can be computer controlled represent the current trend for the next generation of coordinate metrology. Varieties of probe designs are already available and compatible with most different CMMs [6]. The probing system in CMM machines includes stylus and stylus tip which have their own dynamic characteristics during measuring process [7]. The stylus tip contact with the detected surface is the source of signals that will develop the pattern on the working objects. So, the performance of the CMM overall system is very much dictated by the motion precision of the probe tip and its actuator. Therefore, the probe stylus tip is laterally at the center of the CMM operation and a key element of coordinate measurements. The detection probes branched into two main categories; are contact (tactile) probes and non-contact probes. The contact probe gathers data by physically touching of specimen directly, which can be classified into two specific families of manual hard probes and touch trigger probes [8, 9].

Automotive Engine Metrology
Salah H. R. Ali
Copyright © 2017 Pan Stanford Publishing Pte. Ltd.
ISBN 978-981-4669-52-8 (Hardcover), 978-1-315-36484-1 (eBook)
www.panstanford.com

3.1 Types of CMM Probes

3.1.1 Hard Probe

The hard probes are available in a variety of configuration and continue to have a broad application in coordinate metrology. This type of probes is used in conjunction with manual CMMs for low and medium accuracy requirements. Hard probes are simple in use and rugged also, but their repeatability quality depends upon the operator's touch. Because every operator has a different touch when moving and bringing the probe into contact with the feature, therefore this hard-type probe is not commonly used in mass production, which requires a high level of accuracy.

3.1.2 Trigger Probe

A touch trigger probe is the common type used in CMM. It has a precision-built-in and touch-sensitive device that generates an electronic signal through probe tip contact with the specimen surface, which is usually indicated as visual LED light and an audible touch signal. The probe head itself is mounted at the end of one of the CMMs moving axes; it can be rotate automatically, and many different probe stylus tips can be accommodated and attached. The touch trigger probing system function is based on the "three-leg principle" (see Fig. 3.1).

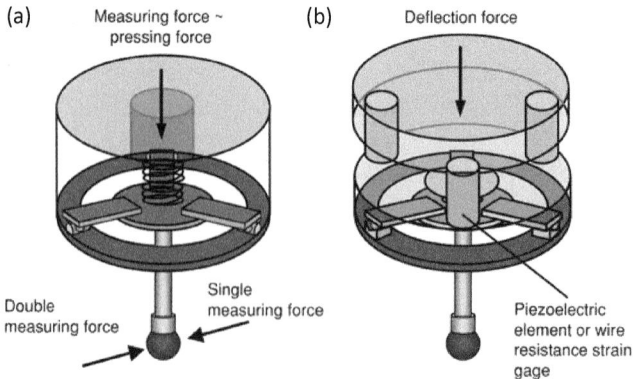

Figure 3.1   Principles of touch trigger 3-D probe [10]: (a) Electro-mechanical contacts; (b) mechanical-electrical converter.

These capabilities make the CMM trigger probe a versatile and flexible data-gathering device. The main disadvantage of this system is due to the variation of probing forces according to the stylus deflection of varying elasticity [10]. CMM equipped with trigger probes eliminate the influence of the operator touch on the quality of measured data compared to hard probe type, because the trigger probe can be fitted on direct computer numerical control (CNC-CMMs) and manual CMMs systems [8, 11].

An additive improvement to the basic touch trigger probe design includes a piezoelectric-based sensing to transmit the deflection of the probe into a constant digital acoustic signal that is recorded by the CMM. This design ensures high rate of measurement accuracy in accordance with the elimination of the stylus bending behavior and probe's internal electromechanical reactions.

Theoretical analysis and experimental studies spotted recently on the touching point as the main source of probe detection errors [1–7, 12–24]. Practically, measurement accuracy is determined by the kinematics accuracy of the CMM probe and a big portion of used machines. Some of these studies observed CMM error combination of coordinate frames as a moving rigid body of the measuring volume and numerical error compensation with different mapping methods effect of geometrical errors lead to accuracy improvement [6, 7]. Considerable research works have been reported to improve the kinematic accuracy of the CMM probe, which is too sophisticated to implement. Few programs focus on changing the CNC program to compensate the probe error [4, 7, 15], waviness deviation of the measured mechanical parts is the desired value owing to many quasi-static systematic errors as inherited intrinsic geometric errors of probe tip, thermally induced distortions of machine probe elements, probe stiffness, touch force and some error sources [1, 2, 5–7]. Software-based error compensation is a method for anticipating the effect combination of all given factors on standard ring accuracy and suitably modifying the conventionally designed probe tip scanning trajectory. A compensation model has been developed to clear parametric study effect on the dynamic measurement errors [21, 23, 24]. Other experimental works have been performed to study the scanning measuring machine. Some of these studies analyzed the effects of scanning speed on the performance of different CMMs to evaluate the measurement performance within special conditions on CMM measurements [23, 24]. However, most of these studies could not separate the performance of the probing system from other error sources through probe scanning. Approaches to derive the dynamic effects of CMM probing systems and techniques of probing compensation have been studied [15, 20–23]. Moreover, the stylus tip of the triggering system is limited by its dynamic root errors that may markedly affect its response characteristics [18, 19]. Therefore, many researchers paid attention to the study of CMM trigger probe characteristics. CMM vibrations in the time domain have been discussed to insure the Monte Carlo methods as a general used tool for understanding dynamic problems resolution [22]. Significant work towards the influence of the object material stiffness, surface shape and its roughness based on measurements of the distance between reference and triggering points in various directions by the rotation of a precise rotary table [2]. That work shows the influence of the specimen material stiffness on pretravel variation depends on measuring force settings as a statistically significant. It has been shown that specimen surface roughness has also significant influence on the touch trigger probe pretravel variation independently on applied measuring force [2]. Another approach for stylus tip radius correction has been based on the use of the part CAD model using straight-line measurement [3]. Some other researchers studied the influence of the actual configuration of the probe length and volume on the accessibility of inner features, such as slots and holes [5]. Dynamic analysis of simple mechanical model of touch trigger probe included the effects of stylus bending and the frictional interaction between the stylus ball and the part surface during applied force has been researched [20]. However, the present work proposes the stylus tip envelop method to define the measured waviness profile and discusses the error resulting from the probe tip angle due to rotation with the real circumference surface contact during scanning at different tip sizes. On the other hand, generating the surface profile of the cylindrical often concavely deviates compared to the ideal straight-line surface; especially, it indicates the necessity of measurement during the measuring process. In the difficult example, the CMM metrology machine is used to measure the amount of deviation in the roundness waviness to determine the measurement errors using suitable measurement strategy similar to probe styli characteristics and a suitable standard ring gauge according to the ISO 10360-6:2009 [7, 25]. More efforts are required for static and dynamic stylus response analyses to classify and evaluate the measurement error sources according to the considered design parameters, especially for the construction of new CMMs [23, 26].

In this chapter, three parameters affecting the measurement errors are considered. The first parameter is the rotation angle of probe stylus tip; the second one is the stylus tip size as unforeseeable root errors. The third parameter is the speed of the CMM touch trigger probe. These measurements are determined accurately using reference artifact including concavely cylindrical surface. Finally, a novel method to verify the dynamic characteristics of the probing system is proposed theoretically and experimentally in order to improve the quality of CMM measurement accuracy.

3.2   Analytical Model

Since the influence of some unforeseeable factors affecting probe inaccuracy could be small, it requires an accurate mathematical model during analysis. Thus, for this investigation, a new two-dimensional-model (2DM) has been used to present the root dynamic errors due to ball tip size of the CMM probe stylus at measurement operations.

3.2.1   CMM Probe Ball Tip Error

Throughout scanning, all touch probes in the CMM coordinate measurement have natural ball tip errors [15, 18, 19]. Supposing a developed 2DM in which the stylus ball is steadily placed in a horizontal position, thus making only the X- and Y-axes translation movement of the stylus. Assuming no stylus tip ball deformation and no surface deformation under the test [18, 19], this 2DM model can be presented as shown in Figs 3.2 and 3.3.

The measurement principle of the proposed system includes contact points 1, 2, and 3 that are indicated on the vertical and horizontal planes of the probe stylus tip with l stylus length and ball tip radius R as shown in Fig. 3.2. Figure 3.3 shows that due to the finite size of the probe stylus ball tip, the contact point on a cylindrical surface will be along the stylus axis, but relatively at some point on the side of the ball where the test surface and the stylus tip ball slope match horizontally. Because the ball does not touch the test artifact specimen along the same stylus slope angle, there will be an error E in the measured length for any measurement point where the test part surface slope (θ) is not 0°. The error E and different possible positions of the ball tip are shown. Case d is at a higher surface slope (180°) and thus has a larger measurement error E, while case a is located at a lower surface slope and has a smaller measurement error E. Therefore, to get the exact location of the point n on sloped surface, an error E, reduced by value ΔY, is made, due to which the position of point t on the stylus tip ball is captured every time. From Fig. 3.3, values ΔY and E can be expressed as follows:

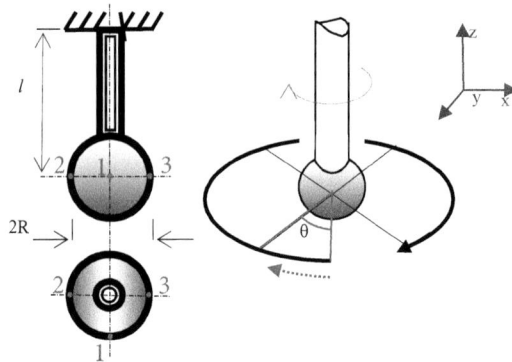

Figure 3.2   Horizontally placed probe stylus ball tip (radius R).

$$\Delta Y = R - r \tag{3.1}$$

$$\Delta Y = R - (R \cos \theta) = R (1 - \cos \theta), \tag{3.2}$$

where the distance between two points t and n indicates the E in the Y direction, while ΔY is the relative distance between points m and t in the Y direction, the large scale of the tip ball in case b. Using the 2DM, it can be stated that measurement error E and ΔY values are made only in the Y-axis and are dependent on the surface slope. In point 2 or 3 (according to Fig. 3.2, where is matching angle θ = 180°), ΔY and error E would be maximal values, while error E and ΔY are equal to 0 only at scanning flat surfaces ($\theta = 0$, 360°) when all points m, n and t are overlaying each other ($m = n = t$), as shown in case a.

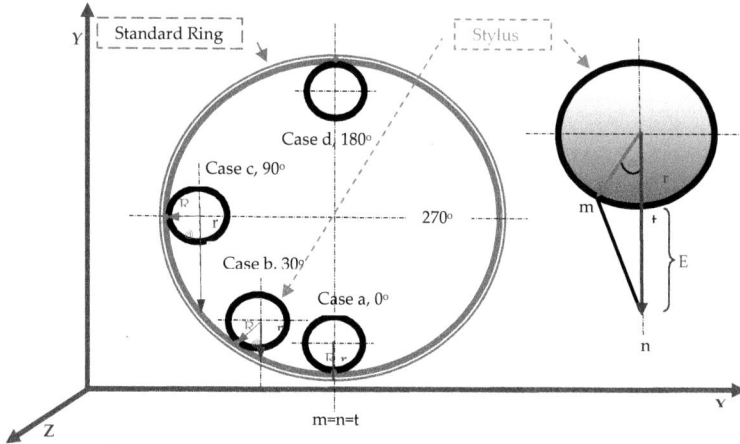

Figure 3.3   Size of error E related to the probe tip ball radius (R) and surface slope degree (θ).

3.2.2   Results of Analytical Model

The root error due to CMM stylus ball tip cannot be neglected. Three different stylus balls with-radii R = 4.0, 2.5, and 1.5 mm have been chosen. A relative mutation in the Y direction (ΔY) of the stylus ball can be observed according to the surface slope degree called the matching angle (θ) (Fig. 3.4).

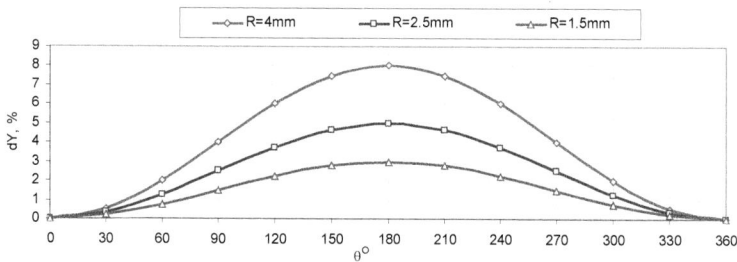

Figure 3.4   Relative error of the ball tip at different surface slope angle.

In the course of the application of accurate analytical 2DM proposed in the measurement of cylindrical parts using CMMs, emerged two types of systematic unforeseeable errors. The first error resulted when increasing surface slope angle while the second error resulted when increasing the radius of the stylus tip ball. Figure 3.4 shows how margin of errors are calculated theoretically and the output of climate surface slope degree (θ) of the probe tip during contact through 360° (using complete cylindrical reference artifact) for different three tips radii. It is observed that the amount of error in the Y direction starting from zero for each probe tip at the beginning of contact at 0°, while incremental increase of inclination angle of a point contact of the tip with the artifact to reach its maximum value of $2R\%$ at 180° and then come back to the decline to reach zero at 360°. It means that rotational motion that occurs during the probe tip scanning due to creeping of the tip at the base of the probe vibrates at the surface coming into contact with the cylindrical parts are also generate another error regularly.

From the theoretical work presented in this chapter, it seems that the results of root errors caused by working mechanism of the touch trigger probe are relatively small. Thus, at the confluence or collection of these errors with other dynamic errors in the measurement, the error values have been tremendously increased as they will be discussed experimentally in the following sections.

3.3 Experimental Work

The structure of the bridge-type-CMM in this study is a typical CMM ductile touch-trigger-scanning probe in the national institute for standards (NIS) at Giza, Egypt. The driving system of Y is located on the right side of bridge. The bridge travels along the Y-guide way, the Y-carriage travels along the X-traverse and the ram together with the probe travels along the X, Y, and Z-directions. ZEISS-Bridge-Type-CMM equipped with PRISMO-VAST touch-trigger probe head under different direct computer control parameters [3, 26]. In the measurement process, on a CNC-CMM the probe is commanded to approach the artifact at a constant speed (positioning velocity) when it comes within the probe approach distance. Experimental work prepared to base on four steps set. This set is prepared to verify the CMM machine according to ISO to be suitable for implementation the schedule of parametric study measurements using experimental set up to achieve the goal of this research.

3.3.1 Verification of CMM Stylus System

Techniques for measure the waviness of the 200 mm standard artifact gauge ring using CMM have been conducted at standard conditions. The ambient environmental conditions during the experiments in laboratory were recorded. The room temperature is maintained within the range 20.0±0.5°C; while the humidity is maintained at 50 ± 2%. The standard ISO 10360-6:2009 [7, 25] specifies the procedure by Gaussian (least-square) for the assessment software, and the probing system has been verified (Table 3.1).

Table 3.1   Results of probing error thought verification scanning

CMM element	Measured radius (mm)	Standard deviation S_D (mm)
Master probe	3.9998	0.0001
Reference sphere	14.9942	0.0001
1st used probe	4.0001	0.0001
2nd used probe	2.5003	0.0001
3rd used probe	1.4992	0.0001

The maximum permissible measurement error (MPE_E) for the used CMM machine evaluated according to ISO 10360-2 [3, 27]; the maximum permissible probing error (MPE_P) and the maximum permissible tangential scanning probing error (MPE_{Tij}) [3, 4, 11, 14, 25] defined as follows:

$$MPE_E = A + L/K, \; MPE_P = \pm 1.0 \; \mu m; \; MPE_{Tij} = \pm 1.90 \; \mu m, \qquad (3.3)$$

where A is a constant value equal to 0.90 µm, L is the measuring length in mm, and K is the CMM manufacture constant equals to 350.

3.3.2 Experimental Procedure

To demonstrate the feasibility of the proposed probe stylus tip size after envelop verification method for corrected measured point determination were carried out on a movable bridge ZEISS PRISMO CMM equipped with VAST scanning probe head of 0.01 µm resolution. The main task of a probing system is to detect whether the tip is in contact with the artifact and to provide a

Table 3.2   Schedule of the parametric study of three work groups

Group	Test no.	Probe stylus specifications			Probe tip scanning speed (mm/s)
		Tip radios (mm)	Mass (g)	Length (mm)	
(a)	1				5
	2	4.0	26.4	63.5	10
	3				15
	4				20
	5				25
	6				30
(b)	7				5
	8	2.5	9.2	53.0	10
	9				15
	10				20
	11				25
	12				30
(c)	13				5
	14	1.5	4.5	33.5	10
	15				15
	16				20
	17				25
	18				30

feedback signal to the CMM control so that all motion from the axes is stopped and the positions of the measured axes can be determined. Experiments were carried out for three different stage types of probe stylus. 3D surfaces of the probe ball tip pretravel are collected using standard test 200 mm ring as an artifact at different scanning speeds with 600 points per scanning track while the traveling speed was 50 mm/s with Gaussian software fitting technique. The probe stylus tip was in contact with the inner circumference surface of the standard gauge artifact to set the origin of the working coordinate system as shown in Fig. 3.5a. Actually, the center of probe tip is the center of measure. The complete schematic of the instrument setup procedures is shown in Fig. 3.5 for three different probe styli including spherical red ruby tips (see Fig. 3.5b). All values of the design parameters of these experiments are mentioned in Table 3.2.

Figure 3.5   Experimental setup using (a) ZEISS-Bridge-Type-CMM, and (b) trigger probe styli with spherical ruby tips.

3.3.3   Parametric Study of Stylus Design

The parameters of study are tabulated based on three different groups a, b, and c as shown in Table 3.2. The measurements were conducted through 90 individual tests, including five repeating times for different scanning speeds, with three different specifications of CMM probe stylus.

3.3.4   Measurement Density

For the CMM probe styli repeatability tests of measurements, a reference standard artifact ring was fixed on the surface of Garnet flat table of CMM as illustrated in Fig. 3.5. The probe head was then rotated through 365° with an interval of 0.01 mm during the circumference of the selected standard gauge ring. The collected data were used to determine the standard deviation for each indexing speed at different probe tip radii using statistical analysis. The variations of form or waviness measurements for the triggered probes were plotted in Figs. 3.6–3.8 to predicate the density of the deviations due to the repeatability of measurements.

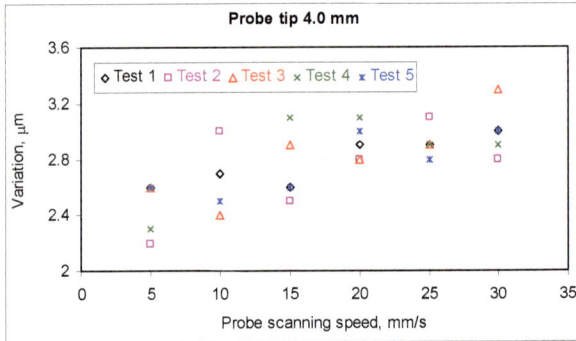

Figure 3.6 Roundness variations for 4.0 mm stylus tip radius size at different scanning speeds.

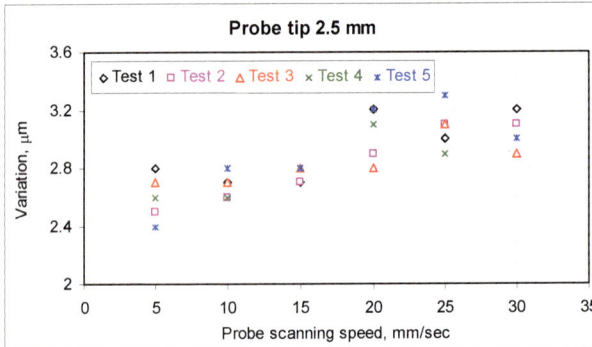

Figure 3.7 Variations of roundness for 2.5 mm stylus tip radius size at different scanning speeds.

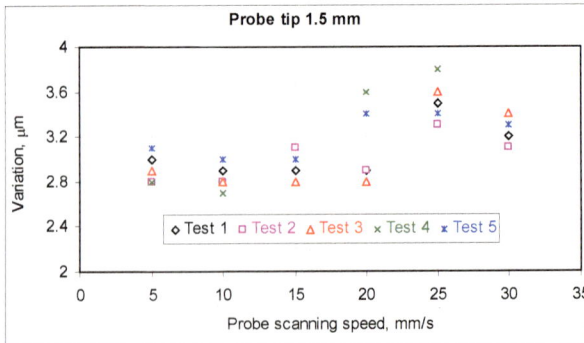

Figure 3.8 Roundness variations at different scanning speeds for stylus tip radius of 1.5 mm.

3.3.4.1 Stylus tip size 4.0 mm

Figure 3.6 presents the results from the measurement repeatability of a touch trigger probe with a 4.0 mm stylus tip radius using the test rig. As can be seen in the figure, the repeatability varies for different approach directions of about 1.1 μm averaged value. Each color (five colors) in the graph in this figure represents a standard deviation for thirty runs at mentioned scanning speeds (Table 3.2). To establish the number of the required repeated runs, test signals were carried out to observe the data variation related to an increasing number of measurements. Based on the results, the number of repeated runs was selected to be five, since no significant difference existed between these measurements more than 0.6 μm.

3.3.4.2 Stylus tip size 2.5 mm

The results in Fig. 3.7 show the density of measured points for roundness waviness variation at different scanning speeds for the same conditions. Once again, the tests were conducted at different approach directions using the same stylus tip of 2.5 mm for the same probe head. These results who stated that the probe repeatability should be within 0.4, 0.2, 0.1, 0.4, 0.4, and 0.3 μm for 5, 10, 15, 20, 25, and 30 mm/s probe tip scanning speed, respectively.

3.3.4.3 Stylus tip size 1.5 mm

For the same conditions, the repeatability error increases dramatically when longer styli are used. Figure 3.8 shows that the repeatability is adversely affected by decreasing stylus tip. Once again, the tests were conducted at different approach directions using the probe stylus tip radius of 1.5 mm with the same CMM probe head. These results showed that the probe repeatability should be within 0.3, 0.3, 0.35, 0.8, 0.4, and 0.3 μm for 5, 10, 15, 20, 25, and 30 mm/s scanning probe tip speeds. It becomes clear from results in Figs. 3.6–3.8 that:

- Any increase in deviations was detected by increasing the speed of measurements for each stylus tip of radius 4.0, 2.5, or 1.5 mm.

- The difference in the deviations between the five repeated measurement values was decreased according to reducing the radius of the stylus tip.
- Minimum deviation belongs to stylus tip with a radius of 2.5 mm at speeds of 10 and 15 mm/s, as well as at the tip radius of 1.5 mm at speeds of 5 and 10 mm/s only.

It is quite clear that the repeatability error is smaller when applying limited stylus tip through the CMM. This suggests that machine errors affect the results of the probing system itself, because the CMM probe system is sensitive to the stylus tip, and it may be due to the hysteresis effects mainly caused by friction at the seating contact are magnified by elastic deflection of the stylus. Obviously, there are some dissimilarities which could be attributed strongly to probe tip bending effects, length and the mass of the stylus beside its tip radius geometry.

3.4 Analysis of the Obtained Uncertainty

Estimation of the uncertainty of waviness measurements for the circler feature is a very complex undertaking, where the uncertainty should be based on many different factors and the procedural steps of experimental and statistical techniques. There are different recognized techniques for determining the uncertainty of coordinate measurements made with CMMs [28].

Regression and variance for experimental data analysis are applied. Table 3.3 shows the statistical determination of the values of M_{AV}, S_D, and U_A. M_{AV} is the mean average value of five repeated test measurements, S_D is the standard deviation, and U_A represents the uncertainty (type A) due to measurement repeatability. The number n is 5 repeating test measurements, and x_i is the measured value in roundness waviness in μm. Figure 3.9 shows the uncertainty of repeatability for the given tip sizes and scanning speeds. It ensures that the largest probe stylus tip has the highest level of uncertainty (type A) within the low measurement speed range from 5 to 15 mm/s, while it generates relative low uncertainty level at 25 mm/s. The maximum level of uncertainty (type A) belongs to the smallest stylus tip, especially at speed of 20 mm/s. The uncertainty has moderate values to the tip radius of 2.5 mm compared to the others at most speeds.

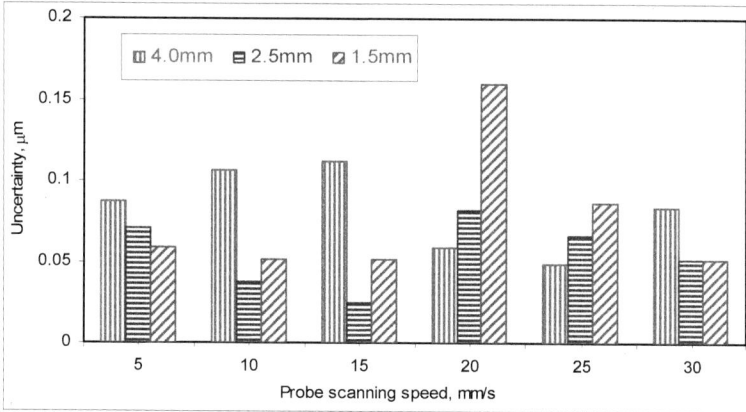

Figure 3.9 Uncertainty (Type A) of roundness form error for probe radii at different scanning speeds.

Table 3.3 Uncertainty (type A) of three probe stylus tip sizes at different scanning speeds

Stylus tip scanning speed (mm/s)	Stylus tip radius (mm)								
	4.0			2.5			1.5		
	M_{AV}	S_D	U_{TypeA}	M_{AV}	S_D	U_{TypeA}	M_{AV}	S_D	U_{TypeA}
5	2.46	0.19	0.09	2.60	0.16	0.07	2.92	0.13	0.06
10	2.62	0.24	0.11	2.68	0.08	0.04	2.84	0.11	0.05
15	2.74	0.25	0.11	2.76	0.05	0.02	2.96	0.11	0.05
20	2.92	0.13	0.06	3.04	0.18	0.08	3.12	0.36	0.16
25	2.92	0.11	0.05	3.08	0.15	0.07	3.52	0.19	0.09
30	3.00	0.19	0.08	3.04	0.11	0.05	3.26	0.11	0.05

3.5 Experimental Results and Discussions

Each measurement of experiment was performed into two compatible sets of the collected data have been carried out. The first set is the selection of probe ball tips includes their specifications of radii, styles mass, styles length and the probing scanning speeds related to cylindrical standard artifact. The second set contains the corrected data set using the CMM built-in algorithm to signify the dynamic errors due to both probe stylus tip sizes and probe scanning speeds in the measurements. Figures 3.10 and 3.13 show the experimental results of some typical selective samples from CMM machine.

3.5.1 Effect of Probe Stylus Tip Size

Figure 3.10 shows the accuracy of waviness profile of response carried out signals of CMM machine in the NIS laboratory for three different probe tip radii at 5 mm/s probe scanning speed both in time domain and in frequency domain using built-in fast Fourier transformer.

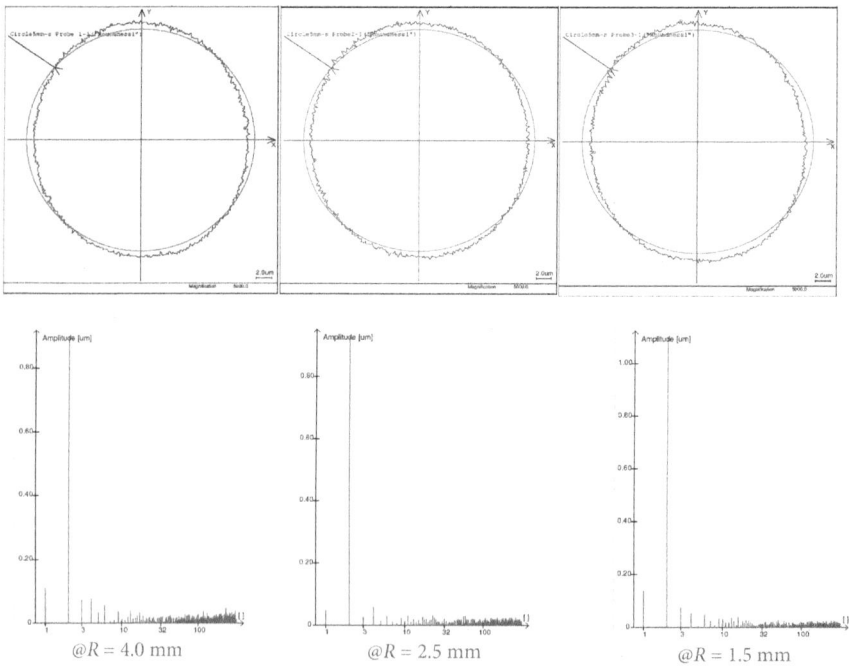

Figure 3.10   Profile waviness accuracy at different probe tip sizes in the time and frequency domains.

Signal analyses illustrate that carried out surface waviness has a significant response at 2 Hz effective frequency for all measured samples. The highest value 1.0952 µm of surface frequency response belongs to the tip radius 1.5 mm, while stylus ball tip radius 2.5 mm recorded the minimum resonance response of 0.9303 μm with a decrement of about 17% of the averaged response for the same natural frequency. Since these detections are carried out of the same path at the same probing speed on the used artifact. Consequently, the frequency response at 2 Hz for the samples clears that this value is a typical natural frequency of the surface detection speed. Therefore, the 17% frequency response variation can be considered as a detection error of stylus characteristics, where the stylus tip radius 2.5 mm is the suitable one for these carried out conditions at low signal noise within the measured frequency band.

The results presented in Fig. 3.10 give a full consensus of the frequency response at the surface wave natural frequency of 2 Hz with the averaged error value for selected trigger probes at the same probing speed of 5 mm/s. It is clear that influencing the stability of the sine wave that appeared at each of the used probes, this may have significance due to the path stability of the measured surface of the used artifact, as an input source, where the difference between frequency response peaks for different probes may represent the level of the impacts due to probe dynamic movements during the surface detection.

Measurement-averaged errors decrease according to any increase in stylus tip radius or probe detection speed as shown in Fig. 3.11. The error values lie within the range from about 3.5 μm to about 2.5 μm, with the greater errors belonging to stylus tip radius 1.5 mm. It can be noticed that the averaged value of the waviness form deviation slope decreased significantly for stylus tip of 1.5 mm measurements at detection speed of 25 mm/s, it may be due to the influence of this speed on the resonance of the CMM machine elements. The averaged errors have significant decreased levels of using tip radius 4.0 mm; it may be due to low dynamic response far away from CMM machine elements critical speeds, especially at 5 mm/s speed.

In other words, Figs 3.11 and 3.12 can help to conclude that small probe tip of 1.5 mm can be better used to diagnose the true state of the surface form of the specimens than that with bigger tip radius of 2.5 and 4.0 mm. This is because the probe tips of the large radii touch a large contact area with the inner surface of the used standard artifact, and vice versa. In this case, the distortion of the measurement result using 1.5 mm probe tip becomes more visible and gives a better estimate of the measured feature profile compared to the results of the 4.0 mm probe tip.

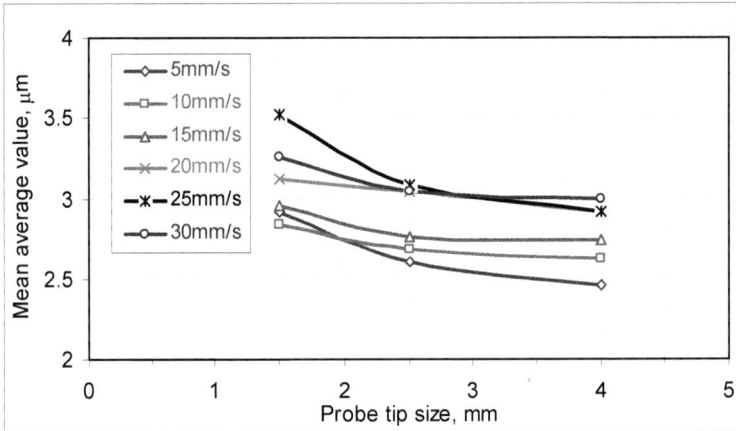

Figure 3.11 Effects of stylus tip radius size on the form error mean values at different probing speeds.

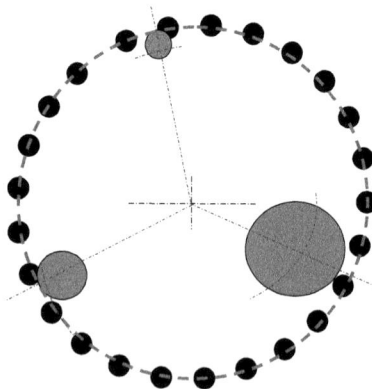

Figure 3.12 Scheme of the probe tips scanning path during measurement.

Figure 3.13 shows six samples on the CMM machine at different probing scanning speeds in the laboratory. It presents the accuracy of the waviness profile of typical measured results at different probe tip scanning speeds of the largest probe tip of 4.0 mm at magnification ratio of 5000:1. The effects of probing speed on the roundness error averaged values for different probe tip radii are detailed in Fig. 3.14.

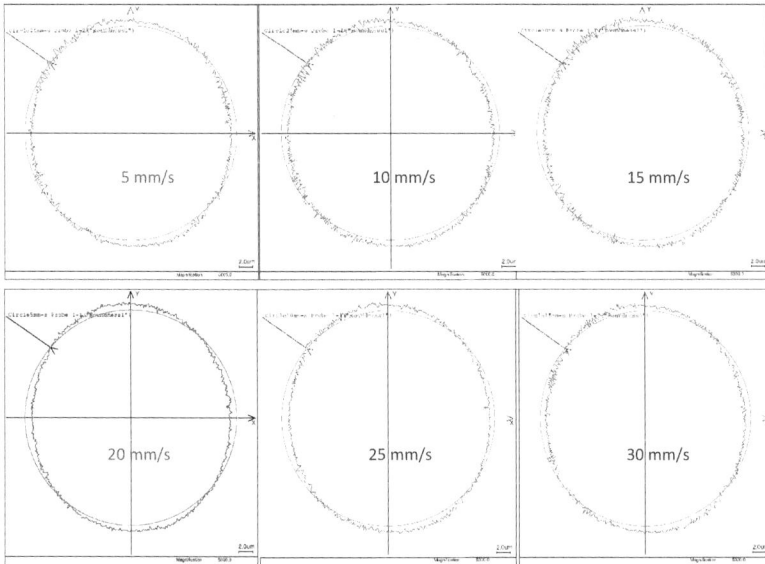

Figure 3.13   Sample response of 4.0 mm probe tip at different scanning speeds.

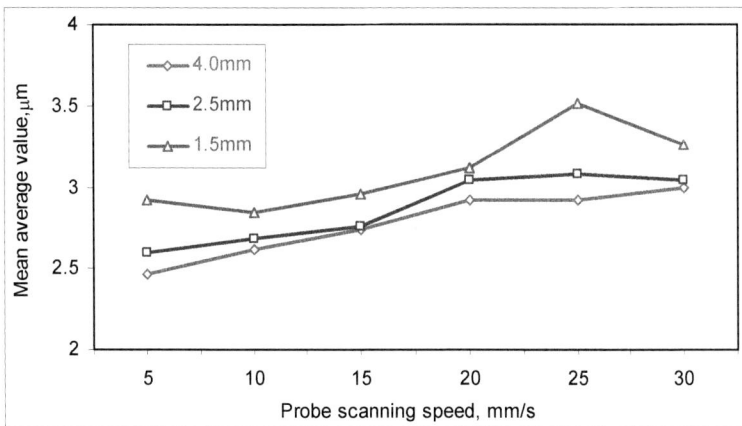

Figure 3.14   Effects of probe scanning speed on the roundness averaged error for different tips.

3.5.2   Effect of Probing Speed

Variations of the probing scanning speed range from 5 to 30 mm/s have a growing rate of averaged errors of sample measurements for all probe tip radii; it may be due to the impact rate of surface asperities with the tip at high scanning speeds (Fig. 3.14). The smallest tip radius generates high rates of averaged error more than that of the other tips due to frequent impacts of relative small asperities at various contact points of the standard artifact surface topography during measuring process, which confirmed its high accuracy of surface roughness diagnosis in accordance with real tip contact area.

The averaged error value of the probe tip 2.5 mm lies between 2.6 and 3.1 µm with a magnitude of about 0.5 µm, while the small probe tip with radius 1.5 mm lies within the range 2.4–3.0 µm with 0.6 µm variation error value. The large tip radius 4.0 mm has an error range between 2.8 to 3.5 µm with varying disbursement of 0.7 µm. However, the maximum deviations of the relative measured errors for both of 1.5 and 4 mm tip radii grow from 10% for scan speed 15 mm/s to reach about 17% corresponds to scan speeds of 5 and 25 mm/s.

3.6   Conclusion

The findings of this study can be drawn into two categories.

(1) From the analytical result analysis, it can be concluded that

- The proposed mathematical 2DM proved to be capable of clearly viewing two different systematic errors with the movement of the probe during scanning. The first error that always occurs due to swerving of the ball tip as it rotates during the scanning process on the measured surface. This error is presented in the Y direction, with zero value where the tip begins to start rotation and reach its maximum value of 2R% at the point of orthogonal axis at 180° and then returns again to zero error at 360°. While, the second systematic error resulted from the increase in the radius of the probe styli ruby ball tips.

(2) From the results analysis carried out, it can conclude that

- Increasing the probe tip radius decreases the averaged measured error signals of surface waviness, which may be attributed to the resulted reduced number of contact points as compared with those of smaller tip on the same artifact surface along the scanning trajectory.
- The amplitude of average error in the waviness profile increased with the probing scanning speed within the range of application. This may be attributed to the vibration excitation of the probe head as it scans the surface. It is evident that the probe stylus tip and probe scanning speed have a significant influence on the accuracy of CMM measurements using the strategy of touch trigger probe independently.
- Based on the results, an easy calibration and correction techniques can be established for probe performance accuracy of CMMs measurements. These techniques can be formulated with relevance of both geometrical surface form and probe stylus dynamic characteristics.

References

1. A. Weckenmann, T. Estler, G. Peggs, and D. McMurtry, Probing Systems in Dimensional Metrology, *CIRP Annals, Manufacturing Technology*, vol. 53, issue 2, pp. 657–684, 2004.

2. A. Woźniak and M. Dobosz, Influence of Measured Objects Parameters on CMM Touch Trigger Probe Accuracy of Probing, Elsevier Inc., *Precision Engineering*, vol. 29, issue 3, pp. 290–297, 2005.

3. A. Woźniak, J. R. R. Mayer, and M. Bałaziński, Stylus Tip Envelop Method: Corrected Measured Point Determination in High Definition Coordinate Metrology. Springer, *The International Journal of Advanced Manufacturing Technology*, vol. 42, pp. 505–514, 2009.

4. Y. Wu, S. Liu, and G. Zhang, Improvement of Coordinate Measuring Machine Probing Accessibility, *Precision Engineering*, vol. 28, pp. 89–94, 2004.

5. J.-J. Park, K. Kwon, and N. Cho, Development of a Coordinate Measuring Machine (CMM) Touch Probe Using a Multi-Axis Force Sensor, *Measurement Science and Technology*, vol. 17, pp. 2380–2386, 2006.

6. G. Hermann, Geometric Error Correction in Coordinate Measurement, *Acta Polytechnica Hungarica*, vol. 4, no.1, pp. 47–62, 2007.

7. H. Schwenke, W. Knapp, H. Haitjema, A. Weckenmann, R. Schmitt, and F. Delbressine, Geometric Error Measurement and Compensation of Machines-an Update, *CIRP Annals—Manufacturing Technology*, vol. 57, pp. 660–675, 2008.

8. D. H. Genest, The Right Probe System Adds Versatility to CMMs, archived at: http://www.qualitydigest.com/jan97/probes.html.

9. M. Dobosz and A. Woźniak, CMM Touch Trigger Probes Testing Using a Reference Axis, *Precision Engineering*, vol. 29, issue 3, pp. 281–289, 2005.

10. Touch Trigger Tactile Sensors, koordinatenmesstechnik, Germany. Website: http://www.koordinatenmesstechnik.de/en/navigation/sensors-for-coordinate-measuring-machines/tactile-sensors/touch-trigger-tactile-sensors.html.

11. Zeiss Calypso Navigator, CMM Operation Instructions and Training Manual. Revision no. 4.0, Carl Zeiss Co., Oberkochen, Germany, 2004.

12. S. H. R. Ali, The Influence of Fitting Algorithm and Scanning Speed on Roundness Error for 50 Mm Standard Ring Measurement Using CMM. *Journal of Metrology & Measurement Systems*, vol. XV, no.1, pp. 31–53, 2008.

13. A. Kasparaitis and A. Sukys, Dynamic Errors of CMM Probes, Diffusion and Defect Data. *Solid State Data. Part B*, ISSN 1012-0394, vol. 113, pp. 477–482, 2006.

14. S. H. R. Ali, H. H. Mohamed, and M. K. Bedewy, Identifying Cylinder Liner Wear Using Precise Coordinate Measurements. *International Journal of Precision Engineering and Manufacturing*, vol. 10, no. 5, pp. 19–25, 2009.

15. Y. C. Lin and W. I. Sun, Probe Radius Compensated by the Multi-Cross Product Method in Free form Surface Measurement with Touch Trigger Probe CMM. Springer, *The International Journal of Advanced Manufacturing Technology*, vol. 21, pp. 902–909, 2003.

16. L. Li, J.-Y. Jung, C.-M. Lee, and W.-J. Chung, Compensation of Probe Radius in Measuring Free-Formed Curves and Surface. *The International Journal of the Korean Society of Precision Engineering*, vol. 4, no. 3, 2003.

17. Z. Xiong and Z. Li, Probe Radius Compensation of Workpiece Localization. *ASME Transactions, J. Manuf. Sci. Eng.*, vol. 125 (1), pp. 101–104, 2003.

18. S. H. R. Ali, Two Dimensional Model of CMM Probing System. *Journal of Automation, Mobile Robotics & Intelligent Systems*, vol. 4, no. 2, pp. 3–7, 2010.

19. M. Sokovic, M. Cedilnik, and J. Kopac, Identification of Scanning Errors Using Touch Trigger Probe Head, *Journal of Achievements in Materials and Manufacturing Engineering*, vol. 20, issues 1–2, pp. 383–386, 2007.

20. W. T. Estler, S. D. Phillips, B. Borchardt, T. Hopp, C. Witzgall, M. Levenson, K. Eberhardt, M. McClain, Y. Shen, and X. Zhang, Error Compensation for CMM Touch Trigger Probes. *Precision Engineering*, vol. 19, issues 2, pp. 85–97, 1996.

21. P. H. Pereira and R. J. Hocken, Characterization and Compensation of Dynamic Errors of a Scanning Coordinate Measuring Machine, *Precision Engineering*, vol. 31, issues 1, pp. 22–32, 2007.

22. J. P. Hessling, Dynamic Metrology: An Approach to Dynamic Evaluation of Linear Time-Invariant Measurement Systems, *Measurement Science and Technology*, vol. 19, no. 8, 2008.

23. A. Farooqui and P. Morse, Methods and Artifacts for Comparison of Scanning CMM Performance, *Transactions of the ASME*, vol. 7, pp. 72–80, 2007.

24. S. Kosarevsky and V. Latypov, Inertia Compensation While Scanning Screw Threads on Coordinate Measurement Machines, *Measurement Science Review*, pp. 68–71, 2010.

25. J. A. Yague, J.-A. Albajez, J. Velazquez, and J.-J. Aguilar, A New Out-of-Machine Calibration Technique for Passive Contact Analog Probes, *Measurement*, vol. 42, pp. 346–357, 2009.

26. J. Zhao, Y. T. Fei, X. H. Chen, and H. T. Wang, Research on High-Speed Measurement Accuracy of Coordinate Measuring Machines, *Journal of Physics: 7th International Symposium on Measurement Technology and Intelligent Instruments, Conference Series*, vol. 13, pp. 167–170, 2005.

27. International Standard (01-12-2009). Geometrical product specifications (GPS)-acceptance and reverification tests for coordinate measuring machines (CMM)-Part 2: CMMs used for Measuring Size, ISO 10360-2.

28. J. Beaman and E. Morse, Experimental Evaluation of Software Estimates of Task Specific Measurement Uncertainty for CMMs, *Precision Engineering*, vol. 34, pp. 28–33, 2010.

Chapter 4

Error Separation of Touch Stylus System and CMM Machine

Coordinate measuring has become the advanced soft metrology technique used in the modern mass production manufacturing. CMM enables determining dimensional and geometrical of surface measurements of the complex objects with relatively high accuracy and precision [1]. The engineering surfaces are viewed as comprising sinusoidal waves of different amplitudes and wavelengths. Measurement of the object surface is a collection of digitized signals of observable physical quantities of error profile. The measurement result exists in the time domain or in the spatial domain using Fourier theory. It comprises sinusoids signals of varying frequencies and amplitudes. This means that by transforming surface profile signals from time domain to the spatial frequency domain, one can clearly identify the amplitude and phase of dominant sinusoidal signals of measured object [2]. Therefore, transformation in spatial domain is very useful, especially in the search for accurate error sources in micro-scale coordinate measurements in a certain frequency range.

Several sources of error in measurement affect the accuracy of CMM result, such as stylus error and machine error. Researchers in the National Metrology Institutes (NMIs) are usually interested in increasing the accuracy of measurement to satisfy higher traceable standard reference and in order to

Automotive Engine Metrology
Salah H. R. Ali
Copyright © 2017 Pan Stanford Publishing Pte. Ltd.
ISBN 978-981-4669-52-8 (Hardcover), 978-1-315-36484-1 (eBook)
www.panstanford.com

improve the quality of engineering produces [3, 4]. Thus, to evaluate the performance of CMM measurements accurately requires determining the values of root errors for each CMM machine separately and in a time adapted to the manufacturing elements. Many factors affecting the CMM machine, operation and stylus errors have been examined and analyzed in detail. To predict and identify these errors and its sources, some investigations on CMMs including typical touch trigger probe errors are very significant and often exceed the errors from different sources [5–8].

Investigations of typical touch trigger probe indicate that probe errors can be generated from different sources [9–11]. Research has also established that the probe errors fluctuate according to the length and mass of the stylus [10–13]. With advanced coordinate soft metrology techniques, became the thorough root errors of CMM machine and stylus system in measurement still need more investigations. Thus, the response of the CMM stylus system during scanning speed needs more specific research analysis to understand the dynamic phenomena in the frequency domain [11]. Figure 4.1 shows a schematic of CMM machine including the probe stylus system.

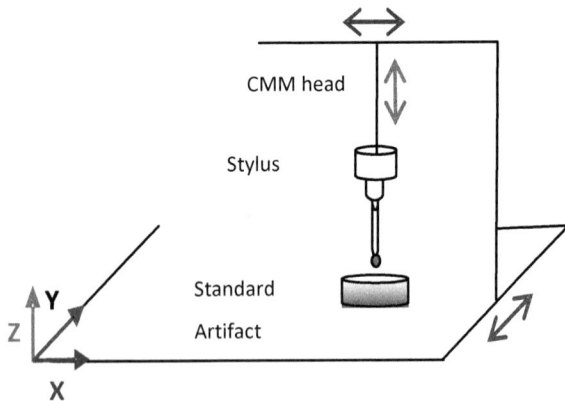

Figure 4.1   Stylus system and CMM machine.

In this chapter, experiment-based Fourier analysis using separate UPR as a spatial frequency response through scanning operation in artifact measurements is done in standard conditions. Empirical equations of root error responses for the stylus system and the machine have been postulated and analyzed. The prediction root errors of both the stylus system and the CMM machine have been determined at the micro-scale response and discussed in detail in this chapter. This work is also very important for correction techniques of CMM development and improves the skills of CMM operator.

4.1 Experimental Work

The Calypso-bridge-type-CMM used in this work is equipped with PRISMO-VAST touch-trigger probe stylus head at the National Institute for Standards (NIS) in Giza, Egypt. The strategy of CMM measurement is controlled to approach the artifact at different positions of scanning speeds. The experiment set is prepared to verify the CMM machine according to ISO including two experimental groups to achieve the goal of this work.

4.1.1 Verification of CMM Machine

The ambient environmental conditions during the experiment in laboratory were recorded. The CMM room temperature is maintained within the range 20 ± 1°C, while humidity is 50 ± 3%. The standard ISO 10360-6 specifies the procedure by Gaussian-least-square fitting for assessing CMM software and probing system have been verified thought verification procedures [14–16]. The measuring coordinate technique has been conducted at standards specific conditions.

4.1.2 Parametric Study of CMM and Stylus Design

The measurement strategy of the study parameters is tabulated based on two different working groups of styli I and II using a standard artifact gauge ring. The measurements were conducted through individual tests including different scanning speeds of stylus from 5 to 30 mm/s with the interval of 5 mm/s as shown in Table 4.1. The geometry of measured surface views many different sinusoidal waves of error amplitudes. Roundness measurement of a cylindrical artifact surface is simply a collection of all frequencies or wavelength sinusoidal components. In other words, the measurement profile of roundness can represent as a superposition of harmonic waves using associated Fourier theory. The fast Fourier analysis serves for the assessment of influences by dominant waviness on the function of a feature of the artifact [2]. A typical separated experimental result of the measurement profile and Fourier analysis of stylus group I at scanning speed of 30 mm/s have been presented in Fig. 4.2. The carried out measurement indicates the waviness profile and Fourier analysis response in time and frequency spaces, respectively, in using Calypso software of the CMM machine at the NIS laboratory.

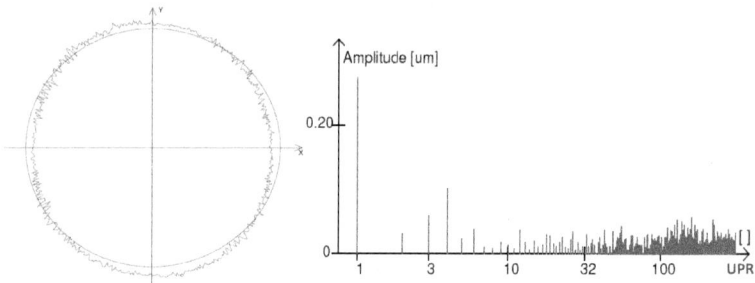

Figure 4.2   Roundness waviness and associated frequency Fourier analysis.

Table 4.1   Two working groups of experiments

		Stylus specifications	
Group no.	Stylus speed, V (mm/s)	Diameter, D (mm)	Mass, M (g)
I	5	3.0	164.8
	10		
	15		
	20		
	25		
	30		
II	5	8.0	186.7
	10		
	15		
	20		
	25		
	30		

The Fourier analysis is the identification method for interferences affecting the measured geometrical profile that occur during form measurement. These interferences can be identified by changing the stylus specifications and stylus speed through the measurement strategy. The stylus scanning speed may affect the recording profile of the geometrical waviness error because of the varying dynamic characteristics of both stylus and CMM machine response. As seen from the Fourier output signal in Fig. 4.2, the sum values of amplitudes of the peaks at different UPR represent the sum error in the measurement. This total value of error has to be more than the resolution value of the CMM instrument. The highest two peak values are indicated parameters for 2DF vibration model to be direct functions of the machine and stylus errors. Consequently, only the highest two peak values in Fig. 4.2 do not represent the total errors. Nevertheless, both of the highest peak values are of original parts from machine and stylus error responses, which will be focused on. Therefore, in this chapter, the author discusses CMM machine- and stylus root errors in detail using a novel experimental analysis technique based on signal separation using mainly the Fourier analysis.

4.2 Analysis of Experimental Results

After verifying the movable bridge CMM-PRISMO machine equipped with VAST scanning stylus with resolution of 0.1 µm, the feasibility of the proposed stylus size was demonstrated to determine the accurate and precise micro-scale measurements. In the experimental, two measurement groups were carried out using standard artifact at different stylus scanning speeds with 300 UPR responses per scanning track, while the traveling speed was 5 mm/s with the Gaussian software fitting technique. Six scanning speeds of stylus have been performed applied for two types of stylus groups. The measurements on CMM machine have been made with different approach speeds of 5, 10, 15, 20, 25, and 30 mm/s. In order to analyze the impact of the machine and stylus responses, the Fourier transformation result of measurement has been separated and presented for different stylus at scanning speeds.

Experimentally, the collected carried out data of the measurement groups are illustrated in Figs. 4.3–4.8. The red signal represents the amplitude of error in measurement at different numbers of UPR response using stylus tip diameter of 3.0 mm with 33.5 mm stylus length and mass of 164.8 g (group I). While the blue signal represents the measurement error at different numbers of UPR response using stylus tip diameter of 8.0 mm with 63.5 mm stylus length and mass of 186.7 g (group II). The peak values of results indicate the stylus system and machine response at different stylus speeds. Therefore, the horizontal linear value of amplitude may be nearly straight line to indicate that the machine continues the error in a white noise form.

The result obtained in Fig. 4.3 gives full consensus on the error values as a dynamic amplitude response of two-trigger-stylus with different UPR at the same scanning speed of 5 mm/s. It clears the stability of the nearly horizontal straight-line behavior which appeared at each of the used two stylus groups (red and blue). This may have significance due to the machine root error average of about 0.03 µm.

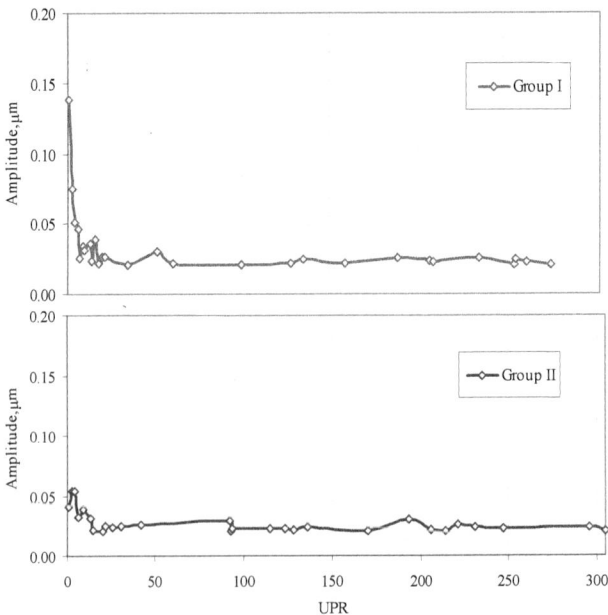

Figure 4.3 Amplitude of error in frequency domain for groups I and II at scanning speed of 5 mm/s.

Moreover, the peak value of dominant amplitude for stylus group I at the first UPR response (red color) was about 0.14 µm, while that value became 0.05 µm for stylus group II at third spatial frequency response (blue color). Stylus group I gives high dominant amplitude error than group II, which may be due to the better dynamic resonance of stylus group II. Therefore, stylus group II gives a better error amplitude than group I at 5 mm/s, which may be due to the better stability of dynamic resonance of stylus group II in measurement at this scanning speed.

Figure 4.4 shows the stability of the nearly horizontal straight line that appeared at each of the used stylus groups, which may be due to the dominant machine root errors of about 0.03 µm at 10 mm/s of stylus scanning speed. Moreover, the maximum value of error amplitude for stylus group I at the first UPR response was about 0.08 µm, while that value became 0.2 µm for stylus group II at the same spatial frequency response. Stylus group II gives worse amplitude error than group I, which may be due to the bad dominant resonance of stylus group II than group I at scanning speed of 10 mm/s.

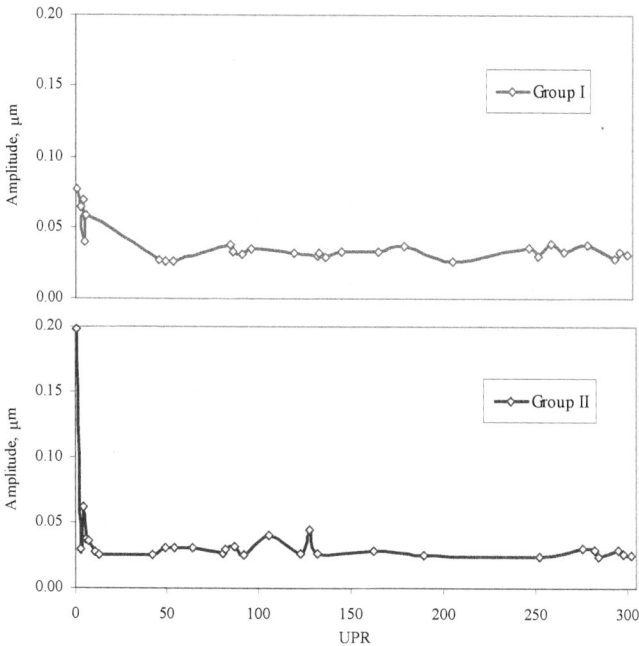

Figure 4.4   Errors at speed of 10 mm/s in spatial frequency response.

Figure 4.5 illustrates the stability of the nearly horizontal straight line that appeared at each of the used stylus groups, which may be due to the machine root errors of about 0.04 μm. Moreover, the peak value of the dominant error amplitude for stylus group I at the first UPR response (red color) is about 0.17 μm, while that value became 0.11 μm for stylus group II at the same frequency response. Therefore, stylus of group I gives better error amplitude than group II stylus at stylus speed of 15 mm/s.

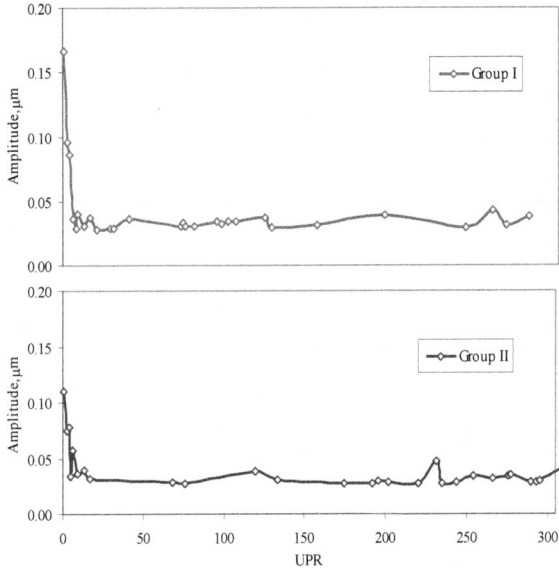

Figure 4.5   Amplitude of measurement at scanning speed of 15 mm/s.

Figure 4.6 shows the stability of the nearly horizontal straight line that appeared at each of the used stylus groups, which may be due to the machine root errors of about 0.045 μm. Moreover, the peak value of error amplitude for stylus group I at the third UPR response (red color) was about 0.07 μm, while that value became dominantly 0.09 μm for stylus group II in the fourth UPR response at 20 mm/s of scanning speed.

Figure 4.7 presents the stability of the nearly horizontal straight line that appeared at each of the used stylus groups, which may be due to the machine root errors of about 0.045 μm. Moreover, the peak value of amplitude for stylus group I at the first UPR response (red color) of about 0.22 μm, while that value became 0.13 μm for stylus group II at the same frequency response and of scanning speed of 25 mm/s.

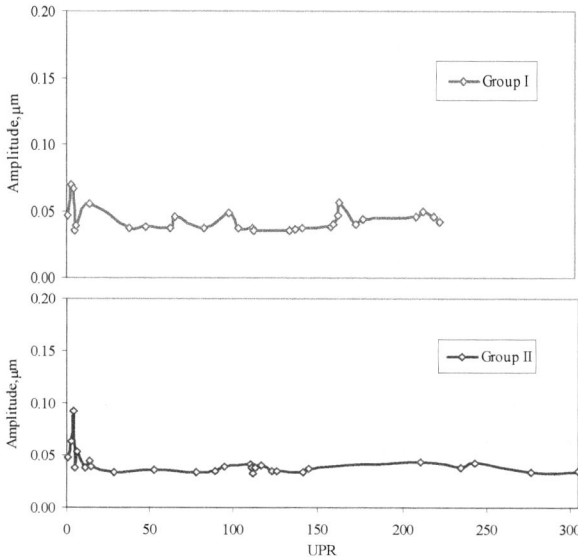

Figure 4.6   The error of measurement at scanning speed of 20 mm/s.

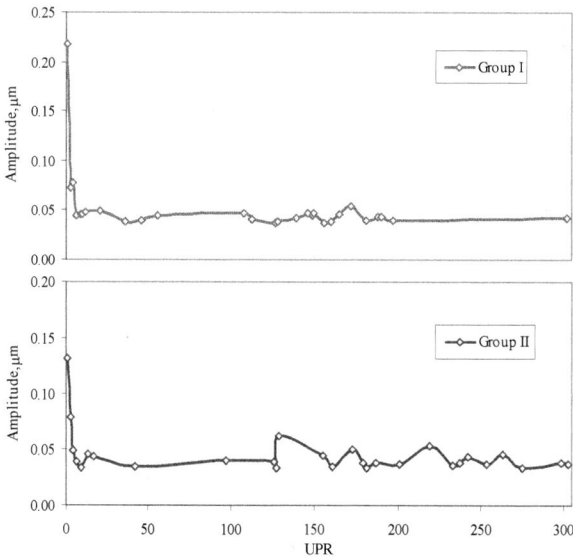

Figure 4.7   Amplitude at scanning speed of 25 mm/s.

The result presented in Fig. 4.8 gives full consensus on the error values as a dynamic amplitude response of two-trigger-stylus with UPR response at the same scanning speed of 30 mm/s. This presentation clears that the effect of the dominant stability of the nearly horizontal straight line that appeared at each of the used stylus groups (red and blue) may be significant due to the machine root errors of nearly 0.05 μm. Moreover, the peak value of amplitude for stylus group I at first UPR response (red color) was about 0.28 μm, while that value became 0.15 μm for stylus group II at the same response (blue color). Stylus first group I gives high amplitude error than group II. Therefore, stylus group II gives a better error amplitude than group I at the same 30 mm/s of scanning speed.

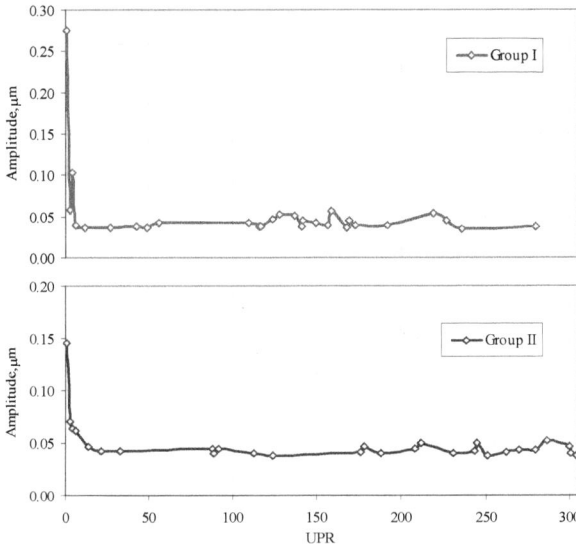

Figure 4.8   Measurement error at scanning speed of 30 mm/s.

Thus, it can be said that stylus group I gives the worse dominant error amplitude than group II at scanning speed of 30 mm/s, which may be due to the given best dominant dynamic resonance of stylus group II in the certain condition. This may be also due to the double stability of dynamic resonance of stylus group II in measurement at this particular speed. Based on the experimental results, the repetition of the amount of the horizontal line with a value ranging from 0.02 up to 0.05 μm is a direct expression of the machine root response at different scanning speeds of stylus from 5 to 30 mm/s. It is can be considered a confirmation of the credibility of validity of the measured results. On the other hand, the values of the stylus root errors ranged from 0.05 to 0.28 μm for the two groups at the certain specific conditions.

4.3 Validation of Experiments

4.3.1 Total Measurement Errors

The validation of the experimental analysis has been presented in Figs. 4.9–4.11. The total error of measurement increases according to the increase in stylus detection speed as shown in Fig. 4.9. The small value of the total error for the first group, which uses the stylus with small size and weight compared to stylus with large size and weight of the second group, is presented. On the other hand, the increase in the value of the total root error (TRER) was attributed to a small variation to the gradual increases of the stylus speed (V) ranging from 0.80 μm at speed 5 mm/s and 1.35 μm at the speed 30 mm/s. This may reflect the impact of stylus dynamic response and reaction through surface scanning of artifact. The error constant value lies empirically within the range of 0.79 to 0.71 μm for the first and second groups, respectively, according to linear numerical trend analyses (Eqs. 4.1 and 4.2).

$$\text{TRER}_{\text{group-I}} = 0.0224V + 0.7920 \qquad (4.1)$$

$$\text{TRER}_{\text{group-II}} = 0.0214V + 0.7064 \qquad (4.2)$$

Moreover, Fig. 4.9 illustrates that the increase of the stylus speed the magnitude of total root error of CMM measurement increases, which may be due to the structure dynamic behavior of stylus design and machine construction. In other words, the smaller diameter and weight of stylus (group I) results give higher error response than that of the large-diameter and -weight stylus (group II).

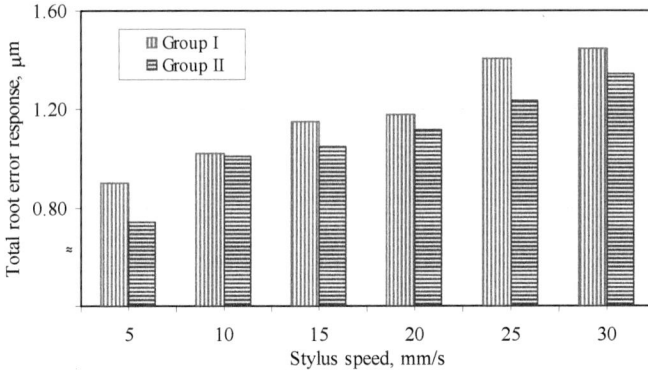

Figure 4.9   Total measurement error after separation at different speeds.

4.3.2   Stylus System Errors

Figure 4.10 shows the error dynamic behavior of CMM stylus systems at different scanning speeds. It illustrates a strong fluctuation according to the variation in the stylus detection speed.

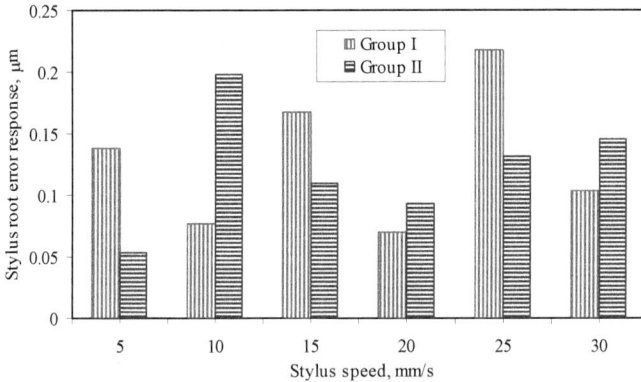

Figure 4.10   Stylus error after separation at different scanning speeds.

The highest root error of the first group is about 0.22 μm at 25 mm/s, while group II error is about 0.20 μm at 10 mm/s. The lowest response of error value of group I is 0.07 μm at 20 mm/s and for the second group about 0.05 μm at 5 mm/s. The error signal values of the stylus take monotonic sinusoidal waveform for group I, and decreased wave of the second group, which are strongly dependent on the stylus detection speed.

The small mass of group I stylus generates relatively high rates of error due to low required energy to oscillate at high speeds, while the second stylus group of higher mass damps the vibration at high detection speeds according to their design parameter considerations. The constant value of stylus error response lies within the range of 0.11 to 0.09 μm in their empirical equations using linear regression type for the first and second groups, respectively (Eqs. 4.3 and 4.4).

$$SRER_{group-I} = 0.0009V + 0.1139 \qquad (4.3)$$

$$SRER_{group-II} = 0.0014V + 0.0976 \qquad (4.4)$$

Error sensitivity is the other important parameter in the dynamics of CMM operation especially for the stylus system. The specific error sensitivity (SRER) of stylus system due to mass has been evaluated, such that the average value of root error response lies within the range of 0.1288 to 0.1217 μm for the first and second groups, respectively. Thus, the relative value of stylus specific error sensitivity (group I to group II) was 120%. This ensures that the measurement using small stylus (group I) is more sensitive than the stylus group II with value of 20% in the certain conditions of this work.

4.3.3 CMM Machine Errors

CMM machine error signals (MRER) of measurements increase gradually according to the increase of the stylus detection speed at certain design and environmental conditions as shown in Fig. 4.11. This may be due to the dynamic behavior of the machine structure. In other words, small diameter and weight of the stylus results in higher detection error amplitude than that of a larger-diameter stylus in the machine error. The machine error constant value lies empirically within the range of 0.028 to 0.024 μm for the first and second groups, respectively, according to numerical linear trend analyses (Eqs. 4.5 and 4.6).

$$MRER_{group-I} = 0.0007V + 0.0281 \qquad (4.5)$$

$$MRER_{group-II} = 0.0008V + 0.0239 \qquad (4.6)$$

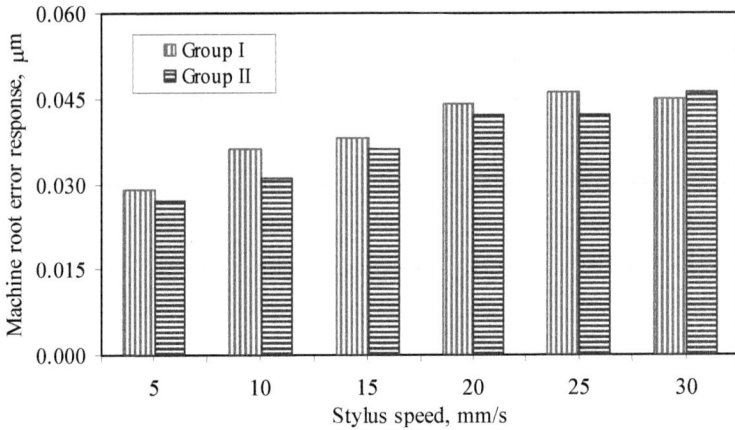

Figure 4.11   CMM machine error after separation versus scanning speeds.

Furthermore, it shows a little increase in the value of the machine root errors of the first set (group I) when using the small-size stylus compared to large-size stylus (group II). The machine root error generated a small gradual increase of stylus speed ranging between 0.028 µm at 5 mm/s and 0.046 µm at 30 mm/s. The highest error belongs to stylus group I due to their small weight. Accordingly, the results of this research can be used for the validation of an experimental investigation to develop the CMM measurement accuracy.

4.3.4   Other Measurement Errors

The stylus system and CMM machine structure errors are not only the errors in the measurements. There are other sources of errors in measurement, such as spindle error, electric motors dynamics, and belt transmutation elasticity and operation- and environmental condition–induced errors. These errors can be determined as the total measured error decreased by stylus system design and the CMM machine structure errors.

4.4   Conclusion

The prediction analysis method of CMM is based on the implementation for the separation of machine structure and stylus design root error responses in measurements based on Fourier analysis at different touch-triggering stylus in the laboratory using the undulations per revolution response. From the present results, the following can be concluded:

(1) The frequency response approach is capable of predicting the root errors of CMM measurements using different stylus systems.

(2) The total average value of root errors of CMM measurement probably lies around 1.08 μm within the presented range of measurement. It may be attributed to the dynamic dominant response of sources of errors such as machine spindle, transmutation elasticity, machine structural properties, and stylus system during measurements.

(3) The stylus with large mass generates low dominant error values at low detection speeds. However, it gives large error amplitude at specified speeds due to dynamic resonance in the certain conditions which should be known to avoid its effect on the quality of measurement. The stylus with small mass produces low error rates, which may be due to the better stabilization of the dynamic response at low detection speeds, while it generates high error values at high speeds due to the worse dynamic resonance and high specific sensitivity of the system. Moreover, the stylus root errors are relatively affected by the stylus tip diameter and stylus scanning speed with value ranging from 0.05 μm to more than 0.21 μm in specific conditions. This may be attributed to the dominant dynamic vibration and sensitivity resonance of the stylus due to its internal design and bending during the scanning of the surface of the measured artifact.

(4) The root errors of CMM machine are dominantly affected by the stylus tip diameter and stylus scanning speed with value ranging from 0.03 to 0.05 μm within the range of measurement application. This may be attributed to the dynamic dominant response of the machine structure during measurement.

(5) The results clearly indicate that the CMM machine root errors are relatively much smaller than the stylus root errors in the present certain conditions.

Acknowledgments

The author appreciates the contributions of *Prof. Dr. M. G. Al-Sherbeny, Cairo University* for his useful comments during final revision of the work.

References

1. S. H. R. Ali, Advanced Measuring Techniques in Dimensional and Surface Metrology, *The 10th International Scientific Conference of Coordinate Measuring Technique, IMEKO-CT14*, Bielsko-Biała, Poland, 23–25 April 2012.

2. B. Muralikrishnan and J. Raja, *Computational Surface and Roundness Metrology*, ISBN 978-1-84800-296-8, Springer, pp. 13–14, 2009.

3. S. H. R. Ali, H. H. Mohamad, and M. K. Bedewy, Identifying Cylinder Liner Wear Using Precise Coordinate Measurements, *International Journal of Precision Engineering and Manufacturing*, vol. 10, no. 5, pp. 19–25, Dec. 2009.

4. S. H. R. Ali, M. K. Bedewy, and Sarwat, Z. Zahwi, Dimensional Inspection of Overhauled Automotive Water-Cooled Diesel Engines, *Proceeding of the International Conference on Production Engineering, Design, and Automatic Control, PEDAC'09*, Alexandria, Egypt, 10–12 Feb. 2009.

5. A. Wozniak and M. Dobosz, Research on Hysterias of Triggering Probes Applied in Coordinate Measurement Machine, *Journal of Metrology & Measurement Systems, Poland*, vol. XII, no. 4, pp. 393–412, 2005.

6. S. H. R. Ali, The Influence of Fitting Algorithm and Scanning Speed on Roundness Error for 50 mm Standard Ring Measurement Using CMM, *Journal of Metrology & Measurement Systems, Poland*, vol. XV, no. 1, pp. 31–53, 2008.

7. W. Huang, Z. Kong, D. Ceglarek, and E. Brahmst, The Analysis of Feature-Based Measurement Error in Coordinate Metrology, *IIE Transactions*, vol. 36, pp. 237–251, 2004.

8. C.-X. J. Feng, A. L. Saal, J. G. Salsbury, A. R. Ness, and G. C. S. Lin, *Design and Analysis of Experiments in CMM Measurement Uncertainty Study, Precision Engineering*, vol. 31, no. 2, pp. 94–101, April 2007.

9. C. Butler, Investigation into the Performance of Probes on Coordinate Measuring Machines, *Industrial Metrology*, vol. 2, pp. 59–70, 1991.

10. S. H. R. Ali, Two Dimensional Model of CMM Probing System, *Journal of Automation, Mobile Robotics and Intelligent Systems*, vol. 4, no. 2, pp. 3–7, 2010.

11. S. H. R. Ali, Probing System in CMM, *The 10th International Scientific Conference of Coordinate Measuring Technique, IMEKO-CT14*, Bielsko-Biała, Poland, 23–25 April 2012.

12. F. M. M. Chan, E. J. Davis, T. G. King, and K. J. Stout, Some Performance Characteristics of A Multi-Axis Touch Trigger Probe, *Measurement Science & Technology*, vol. 8, pp. 37–48, 1997.

13. A. Wozniak and M. Dobosz, Factors Influencing Probing Accuracy of a Coordinate Measuring Machine, *IEEE Transactions on Instrumentation and Measurement*, vol. 54, pp. 2540–2548, 2005.

14. H. Schwenke, W. Knapp, H. Haitjema, A. Weckenmann, R. Schmitt, and F. Delbressine, Geometric Error Measurement and Compensation of Machines-an Update, *CIRP Annals-Manufacturing Technology*, vol. 57, pp. 660–675, 2008.

15. J.-A. Yagüe, J.-A. Albajez, J. Velázquez, and J.-J. Aguilar, A New Out-of-Machine Calibration Technique for Passive Contact Analog Probes, *Measurement*, vol. 42, pp. 346–357, 2009.

16. A. Nafi, J. R. R. Mayer, and A. Wozniak, Novel CMM-based Implementation of the Multi-Step Method for the Separation of Machine and Probe Errors, *Precision Engineering*, vol. 35, pp. 318–328, 2011.

Chapter 5

Measurement Strategies of CMM Accuracy

5.1 Introduction

Nowadays, it cannot be over-emphasized that the world development is according to ultra-high precision technologies. In both manufacturing and measuring technology, an ongoing trend for higher accuracies can be observed. The favor for these technologies is mainly for the engineering metrology. Engineering coordinate metrology is an important branch of quality assurance [1]. Therefore, CMM machines are installed in many large-scale industrial factories, medical laboratories, and scientific research centers, as well as airspace, airplane, and automotive industries. CMM machines are used to measure the surface quality of machine elements and spare parts of the cylinder, piston, gear, and fuel injector nozzles.

Since the required tolerances for manufacturing continuously become smaller, whereas the complexity of work pieces increases, capable measurement techniques have to be applied in order to achieve accurate results with sufficient precision. The final accuracy of a work piece measurement quality is influenced by many different factors [1–3]. The main factors that influence the results of coordinate metrology are presented in Fig. 5.1.

Automotive Engine Metrology
Salah H. R. Ali
Copyright © 2017 Pan Stanford Publishing Pte. Ltd.
ISBN 978-981-4669-52-8 (Hardcover), 978-1-315-36484-1 (eBook)
www.panstanford.com

Figure 5.1 Parameters influencing the results of coordinate measurements.

The resultant measurement quality of CMM is limited by deviations and some uncertainties. The measurement deviations in coordinate metrology can be related to the operator performance quality, environmental interaction, work piece finishing, and CMM accuracy. It can be assumed that some influence factors of operator behavior and CMM software accuracy have effective reactions on the measurement quality factors [1].

An error compensation issue of a CMM has been studied in the context of several factors using different versions of machine software [4, 5]. Earlier researchers have focused on active error compensation of the deterministic error components based on simple models [6–10]. Both of the above error factors have not been studied for their effect on the measurement quality for the same CMM.

In industrial production, the true surface can never be known exactly. Therefore, an approximation of the surface is known based on coordinate points using a finite sampling method. The CMM including special software is aimed to detect the geometry of the surface. The CMM fitting software uses the coordinate data to determine a part's location, orientation, form, and deviation of roundness. The fitting algorithms of testing and evaluation for CMM have been in existence since receive a new CMM machine at NIS against a reference algorithm to include a more extensive test program. The advanced coordinate-measuring machine model PRISMO Navigation has been delivered to the NIS in January 2006 to offer more accurate measurement services.

In this chapter, the CMM fitting software techniques for cylinder roundness measurements, equipped with nine different probe speeds using three detection circles, are studied dynamically and discussed in details. The tests have been performed to examine the problem of how to generate reference data sets of measurement strategy for cylinder circle surface at NIS. These reference data sets are presented to get an optimal strategy at dynamic performance for a CMM machine. The objective is to eliminate the repeatable error in turning operations on CMM machines. The goal is to reduce costs according to operation time and improve figure accuracy of visible measurement in a production environment.

Consequently, some CMM error formulae have been postulated to correlate the roundness measurement errors with the probe scanning speed factor for different fitting algorithms within the application range. The objective of the research prepared is to help the CMM operator in developing a methodology for precision assembly as well as error compensation methods to improve the overall system accuracy. This study is very important also for the CMM designer to develop new precision machines.

5.2   Background and Motivation

In the last few years, the technology of dimensional engineering metrology has been developed specially for large surface instrument manufacturing. In general, the industry has been somewhat reluctant to invest in fitting algorithms software. The success of any fitting algorithm application is derived through the abilities of its software fitting performance and characterization. Many of these software systems employ windows-based software to give the CMM user/operator a highly intuitive visual compatibility with logical, menu-driven functions having comprehensive help facilities for operator support.

There are three types of tolerances: the form, position, and size tolerance. The form tolerance is the largest possible deviation of an element form. Deviation of work piece form is the value of the deviation of the real form to its nominal design form. Irregularities of surface can be decomposed into form, waviness, and roughness. Waviness is the important variable of the geometric dimension and tolerance in engineering metrology. Waviness includes five different effective parameters: straightness, flatness, roundness, cylindricity, and surface profile. Roundness is an essential parameter for any circle and cylinder measurements. To measure roundness, it should include a rotational factor to the measurement, conversely, diametric measurement.

Roundness measuring instruments tend to be using one of two techniques: Talyrond or CMM methodology. Historically in 1954, the rotating pick-up version of the instrument was first made commercially available; this was termed "Talyrond-1," which developed later. Another way to measure surface roundness is to use a coordinate-measuring machine. A standard CMM has three accurate orthogonal axes and is equipped with a touch-trigger probe. The probe is brought into contact with the component being measured at a recorded position. A number of points are taken around the component and these are then combined in computer software to determine the roundness of the component. Typically, the number of data points is very small because of the time taken to collect them. As a result, the accuracy of such measurements is compromised to evaluate the roundness.

5.2.1 Types of Errors

The purpose of CMM software system and operator skill is to determine the final dimensions of the work piece and to provide information about the presented errors in the measurement strategy. Machined surfaces cannot have perfect forms due to various error sources, interaction of machining processes, quality, and measurements' strategic accuracy. Consequently, in this study, the sources of such surface imperfection or errors will be analyzed to cover two main particular error types. The first error is based on two sub-errors of *form error* and *measurement error*. The *form error* conveys the idea that the work piece has not perfectly the shape of their nominal geometry. Even if the CMM machine were somehow perfect, the point measurements would generally still deviate from the nominally perfect shape. Some form of errors can be expected in many engineering work pieces. The *measurement error* arises when data points are collected on the surface of an object. Sources of error based on CMM adaptation (axial bends in some hardware, probe system imperfections, fixture, etc.), and measurement environment (temperature, vibration, etc.) lead to some inaccuracies in the measured points.

The second main error is called *human error*, since the human interaction is yet another big source of error (sometime even the largest source of error). The *human error* arises when the measurement machine operator (metrologist) selects impossible measurement strategy parameters for requirements of an object. Therefore, the CMM operator behavior has a significant effect on the measurement errors. These two main types of errors with some others exist in all real-world measurement scenarios. Therefore, the main purpose of the study is projected to study the influence of CMM fitting algorithms through nine different probe scanning speeds for three different transverse circle locations of carrying out signals, to

(a) develop the CMM software using closed-loop control and to reduce the size of measurement uncertainty;
(b) increase operator skills and reduce the operation lost time and cost, to avoid processing mistakes of software strategy applications.

5.2.2 Fitting Algorithm

The job of the CMM fitting algorithm software is to process the data in such a way that it will be useful to the user. The algorithm testing and evaluation program for coordinate-measuring machines has been studied since 1993. There are two main types of circle fitting software algorithms used in CMM called *Gauss* and *Chebyshev* (*Tschebycheff*) [5]. *Karl Friedrich Gauss* (1777–1855), one of the most renowned mathematicians, befits his elegant "least squares" approach attempt to minimize the average error. *Panutij Chebyshev* (1821–1894), with his minimum distance approach, addressed the bumps smoothed over by *Gauss'* attempts to minimize the maximum error. Moreover, other new algorithm types of CMM software draw in measurement strategy applications.

5.3 Experimental Work

5.3.1 General

In this work, the evaluation of the CMM PRISMO navigator software program through sample carrying out signals has been performed. The evaluation processes include three basic components of the instrumentation system: a data generator, reference algorithm, and a comparator to analyze and interpret the results. The CMM has six fitting categories. The machine software algorithm and probe scanning speed were selected and primary tested in the recommended environmental conditions at NIS laboratory. An eccentric work piece seat base of granite was finely cleaned and located on the CMM test position. The CMM machine was turned on to check the electric power switches and pneumatic pressure, where a styles probe of the long type has been selected and calibrated according to the machine working manual. The performance of the CMM accuracy in scanning measuring mode was verified and accepted within standard specification according to ISO 10360 [11, 12]. An inspection feature consists of one or more surface elements, like a cylinder in this case to find the associated tolerance. In order to determine the accuracy of the approximation fitting algorithm, a geometrical model type casing for future tests was created. Many aspects of roundness error measurement strategy have been taken into account according to standard procedures.

5.3.2 Dynamic Calibration of Stylus System

Dynamic calibration of the CMM stylus system is very important, especially in the field of study CMM accuracy [12–14]. The standard measurement methods of both probing error and scanning probing error using reference sphere. The diameter of the reference standard test sphere should be between 10 and 50 mm with certification for form and diameter. To determine the probing error, one must probe 25 recommended points on the reference test sphere surface. To determine the CMM scanning probing error, one must scan four recommended scanning lines on the surface of test sphere and compute the Gaussian center point of the sphere using all measured points of all four scan lines. Before making measurements with the CMM in the cylinder, the CMM was calibrated using master probe for evaluate standard sphere and using standard sphere for evaluate used probe. The output standard deviation (SD) and CMM test element specification are presented in Table 5.1.

Table 5.1   Output data of CMM probes and sphere

CMM element	Measured radius (mm)	SD (mm)
Master probe	4.0000	0.0001
Reference sphere	14.9942	0.0001
Used probe	4.0000	0.0001

The CMM has limited specific values of as follows:

$$\text{MPE}_E = A + L/K, \mu m,$$

where MPE_E is the maximum permissible measurement error, A is the constant machine uncertainty equal to 0.9 µm, K is the length constant or slope of line equal to 350, and L is the length measurement in mm.

$$\text{MPE}_P = \pm 1.00 \ \mu m \text{ and } \text{MPE}_{Tij} = \pm 1.90 \ \mu m,$$

where MPE_P is the maximum permissible probing error and MPE_{Tij} is the maximum permissible error when measuring a part by using scanning mode which called maximum permissible scanning probing error.

5.3.3   Test Procedure

After CMM adjustment and calibration, a finely finished steel work piece as a ring block has been prepared for the test. The work piece has an outer/inner diameter 82/50 mm, height of 10.2 mm (Fig. 5.2). According to the following plan, measurements have been carried out at three different transverse sections on the work piece inner circles at locations of 4, 4.25, and 4.5 mm from the top to detect any roundness error of the surface. The measured sample errors were obtained for nine probe scanning speeds 5, 10, 15, 20, 25, 30, 35, 40, and 45 mm/s during 360° angle range trace of the standard ring, respectively, while the CMM traveling speed was constant of 15 mm/s and the number of scan fitting points also was constant with about 1633 ± 2 points during measurement tests at temperature condition of $20 \pm 0.5°C$. Each measurement point has 10 times repetitions for the same three transverse circle $(x, y,$ and $z)$ positions.

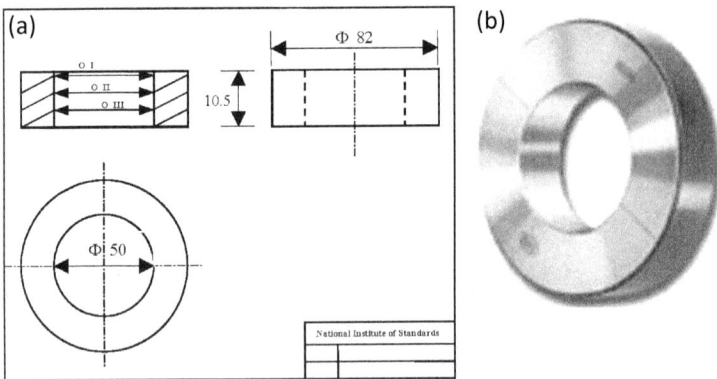

Figure 5.2   The tested standard ring block. (a) A simple drawing of the cylinder. (b) The selected steel ring.

The new CMM software has six fitting algorithm types for all measurement applications as follows:

(a) Least square (*Gauss* criterion method), LSQ

(b) Minimum element (*Chebyshev* criterion method), ME

(c) Minimum circumscribed element (calculation method), MCE

(d) Maximum inscribed element (calculation method), MIE

(e) Inner tangential element (calculation method), ITE

(f) Outer tangential element (calculation method), OTE

Actually, in the roundness measurement application, the ITE fitting technique is equal to MCE technique and OTE fitting technique is equal to MIE technique. Therefore, in the selected three circles, roundness has been determined at each probe scanning speed, where the CMM PRISMO navigator has been selected to the above first four fitting algorithms LSQ, ME, MCE, and MIE only. Determination of roundness measured errors has been included in 1080 experimental measuring tests to differentiate between evaluation qualities of the different measurement strategies.

5.4   Result Presentation and Discussion

The PRISOM CMM data fitting using the different ways yields a drastically different resulting geometry. Analyses of roundness error of the four fitting techniques are given in details. However, a question arises which method is suitable to choose and what criterion should be taken at which probe scanning speed. The density of measured points is presented in Figs. 5.3, 5.5, 5.7, 5.9, 5.11, 5.13, 5.15, 5.17, and 5.19. The results obtained are reduced and presented in a more practical and explicit form in Figs. 5.4, 5.6, 5.8, 5.10, 5.12, 5.14, 5.16, 5.18, and 5.20. The roundness error results as functions of the probe scanning speed and fitting technique are given as follows.

5.4.1   Probe Scanning Speed 5 mm/s

The presentation of 120 test results in Fig. 5.3 shows the density of measured points for roundness error using different four fitting algorithms of three transverse circles at probe scanning speed of 5 mm/s. Figure 5.4 shows the average variation of roundness errors of different types of fitting algorithms for three circle cases I, II, and III. The analysis of the given results indicates the following:

- Detection *circle I* measurements have a roundness error range of 0.23 µm from 2.20 to 1.97 µm, while measurements of *circle III* have roundness error limits of 2.11 and 1.81 µm within an error range of 0.30 µm.
- *Circle II* measurements have the highest roundness error 2.65 µm for the MCE fitting method, and the lowest error of 2.38 µm for the ME fitting method. Consequently, *circle II* have the highest roundness error 2.55 µm for the LSQ fitting method.
- According to the application of the fitting technique to all measuring circles, the evaluated error difference between circle measurements as representing values to the fitting method quality, have 0.44 µm of LSQ (2.55 and 2.11 µm) and have 0.41 µm of ME (2.38, 1.97 µm), while MCE (2.65 and 2.1 µm) and MIE have the values 2.59 and 2.01 µm at a maximum difference of 0.55 and 0.58 µm, respectively.

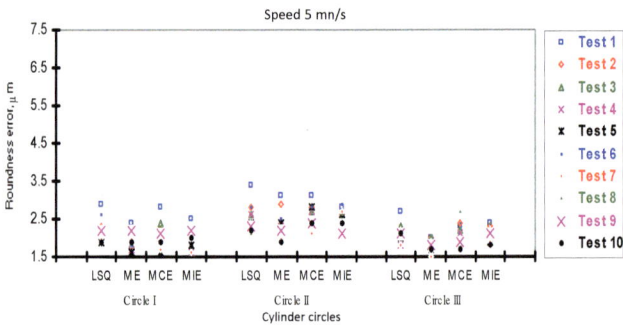

Figure 5.3 Roundness errors variation of fitting algorithms for three detection circles at 5 mm/s.

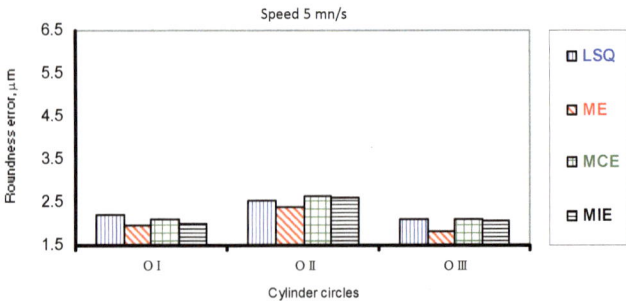

Figure 5.4 Measuring errors average of different fitting algorithms at 5 mm/s in the ring circles.

5.4.2   Probe Scanning Speed 10 mm/s

Figure 5.5 shows the density of 120 measured tests for roundness error using different four fitting algorithms of three different circles at probe scanning speed of 10 mm/s. Figure 5.6 shows the variation of roundness errors of different types of fitting algorithms for three circle cases I, II, and III. The analysis of the results indicates the following

- *Circle I* measurements have the highest roundness error 1.93 μm for the LSQ fitting method, and the lowest error of 1.68 μm for the ME fitting method. Consequently, *circle I* has a measuring error range of 0.25 μm.
- Detection *circle II* measurements have a roundness error range of 0.33 μm from 2.01 using MIE fitting method to 1.68 μm using ME fitting method, while measurements of *circle III* have roundness error limits of 2.24 using MIE fitting method and 1.74 μm using ME fitting method within an error range of 0.50 μm.
- According to the application of the fitting technique to all measuring circles, the evaluated error indicated clear difference between circle measurements as representing values to the fitting method quality has 0.43 and 0.09 μm for MIE (2.24 and 1.81 μm) and LSQ (1.98 and 1.89 μm), while ME (1.74, 1.68 μm) and MCE have the values 1.83 and 1.80 μm at a minimum difference of 0.06 and 0.03 μm, respectively.

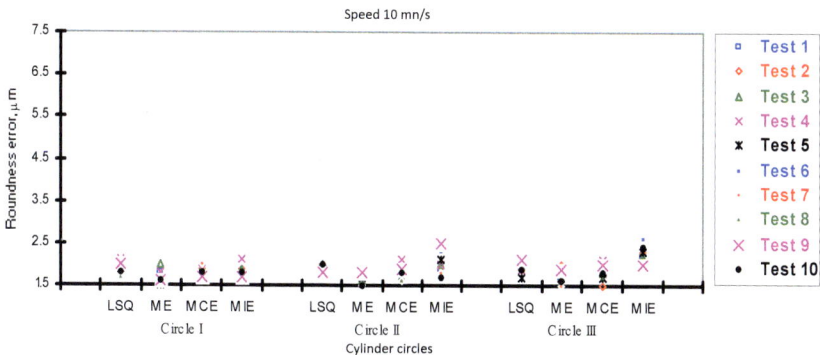

Figure 5.5   Measuring errors variation of fitting algorithms for three circles at scanning speed of 5 mm/s.

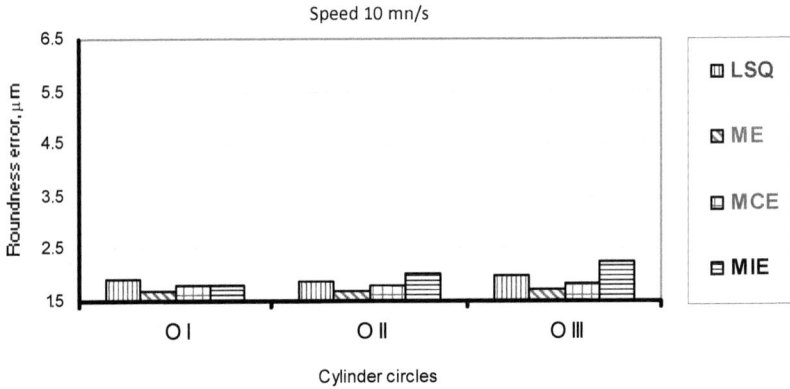

Figure 5.6 Roundness errors of fitting algorithms at 10 mm/s for three detection circles.

5.4.3 Probe Scanning Speed 15 mm/s

The results in Fig. 5.7 show the density of measured tests for roundness error using different four fitting algorithms of three circles at probe scanning speed of 15 mm/s. Figure 5.8 shows the variation of roundness errors of different types of fitting algorithms for three circle cases I, II, and III. The analysis of the results indicates the following:

- *Circle I* measurements have the lowest roundness error 3.98 μm for the LSQ fitting method, and the lowest error of 3.80 μm for the ME fitting method. Consequently, *circle I* has a measuring error range of 0.18 μm.
- Detection *circle II* measurements have a roundness error range of 0.16 μm from 5.33 to 5.17 μm, while measurements of *circle III* have roundness error limits of 5.44 and 5.33 μm within an error range of 0.11 μm.
- According to the application of the fitting technique to measuring circles, the evaluated error difference between circle measurements as representing values to the fitting method quality, has 1.60 and 1.49 μm for MCE (5.40 and 3.80 μm) and MIE (5.33 and 3.84 μm), while LSQ (5.44, 3.98 μm) and ME have the values 5.33 and 3.8 μm at a minimum difference of 1.46 and 1.43 μm, respectively.

- The measuring error range has significant variation at 15 mm/s compared to 5 and 10 mm/s testing speed, may be due to probe response at resonance traveling speed.

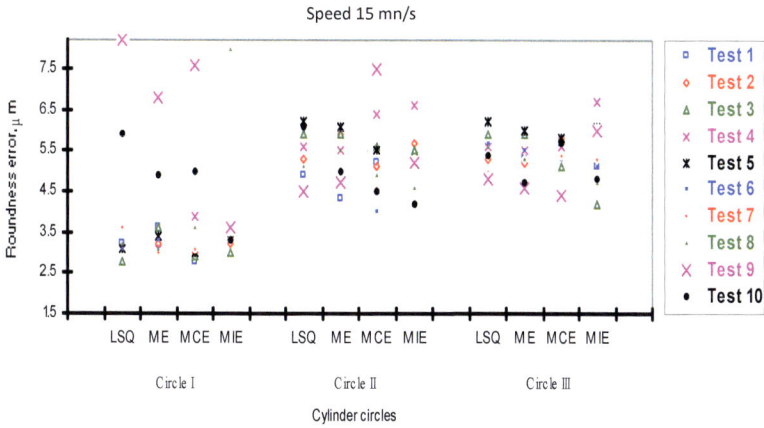

Figure 5.7   Measuring errors variation of fitting algorithms for three detection circles at 5 mm/s.

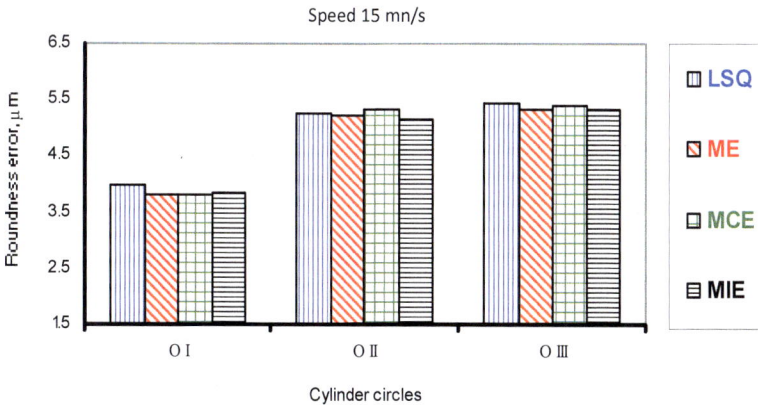

Figure 5.8   Roundness errors of fitting algorithms at 15 mm/s for three detection circles.

5.4.4   Probe Scanning Speed 20 mm/s

Figure 5.9 shows the density of measured points for roundness error using different four fitting algorithms of three transverse circles at probe scanning speed of 20 mm/s. Roundness errors as functions of the fitting techniques are shown in Fig. 5.10. The analysis of the results indicates the following:

- Measurements of *circle I* have a maximum error of 4.10 μm for MIE fittings and a maximum error of 4.98 μm for LSQ response of an error range of 0.88 μm.
- Roundness error ranges of both *circle II and III* have 0.40 μm (5.35 and 4.95 μm) and 0.28 μm (5.35 and 5.07 μm).
- Error difference related to the fitting technique for three measured circles has a maximum value of 1.25 μm (5.35 and 4.1 μm) for MIE fitting response and a minimum value for the same measurements 0.15 μm (5.10 and 4.95 μm) for LSQ and 0.97 μm for ME fittings.

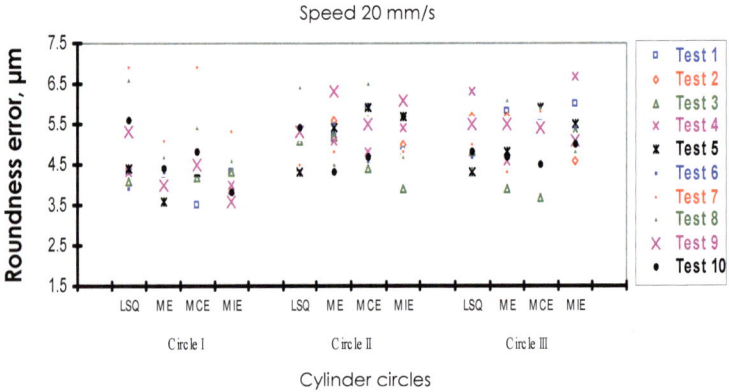

Figure 5.9   Roundness errors variation of fitting algorithms for three detection circles at 5 mm/s.

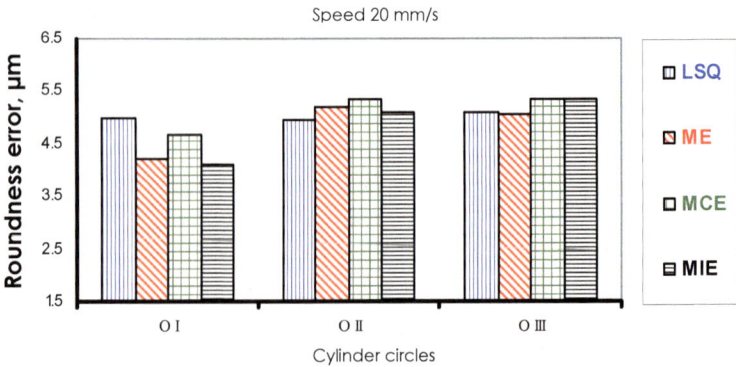

Figure 5.10   Measuring errors of fitting algorithms at 20 mm/s for three detection circles.

5.4.5   Probe Scanning Speed 25 mm/s

The results in Fig. 5.11 show density of measured points for roundness error using different four fitting algorithms of three transverse circles at probe scanning speed of 25 mm/s. Roundness errors as functions of the fitting techniques are shown in Fig. 5.12. The analysis of the results indicates the following:

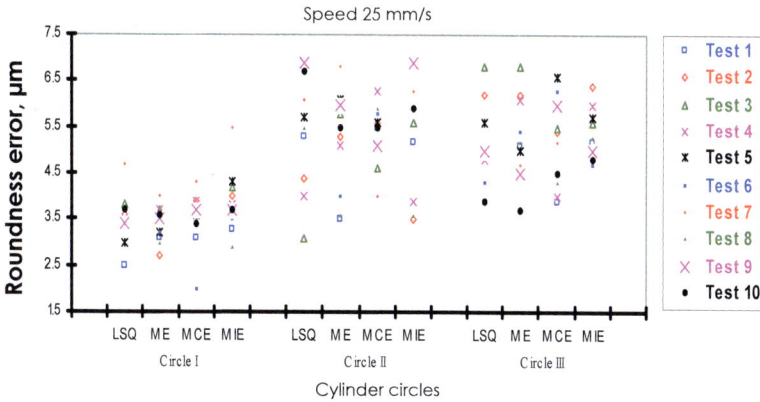

Figure 5.11   Measuring errors variation of fitting algorithms for three circles at scanning speed of 5 mm/s.

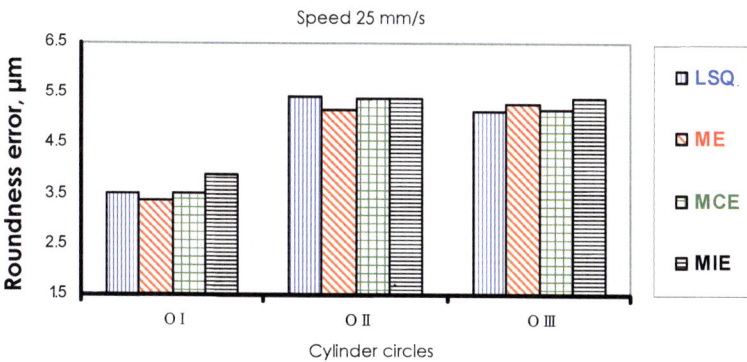

Figure 5.12   Roundness errors of fitting algorithms at 25 mm/s for three detection circles.

- Measurements of *circle I* have a minimum error of 3.37 µm for ME fittings and a maximum error of 3.89 µm for ITE response of an error range of 0.52 µm.

- Roundness error ranges of both *circle II and III* have 0.27 μm (5.44 and 5.17 μm) and 0.21 μm (5.38 and 5.17 μm).
- Error difference related to the fitting technique for all three measured circles has a maximum value of 1.9 μm (3.54 and 5.44 μm) and (3.37 and 5.27) for both LSQ and ME fitting methods and a minimum value for the same measurements 1.49 μm (5.4 and 3.89 μm) for MIE and 1.88 μm for MCE fittings.

5.4.6   Probe Scanning 30 mm/s

Figure 5.13 shows the density of measured tests for roundness error using different four fitting algorithms of three different circles at probe scanning speed of 30 mm/s. Measured error results of roundness for fitting probe 30 mm/s are given in Fig. 5.14 for the detection circles. From the signal analysis, the following can be noticed:

- *Circle I* measurements have the lowest roundness error 3.21 μm for the ME fitting method, and the highest error of 3.56 μm for the MCE fitting method. Consequently, *circle I* has a measuring error range of 0.35 μm.
- Detection *circle II* measurements have a roundness error range of 0.70 μm from 5.87 of MCE fitting method to 5.17 μm of ME fitting method, while measurements of *circle III* have roundness error limits of 5.44 of MCE fitting method and 4.91 μm of LSQ fitting method within an error range of 0.53 μm.

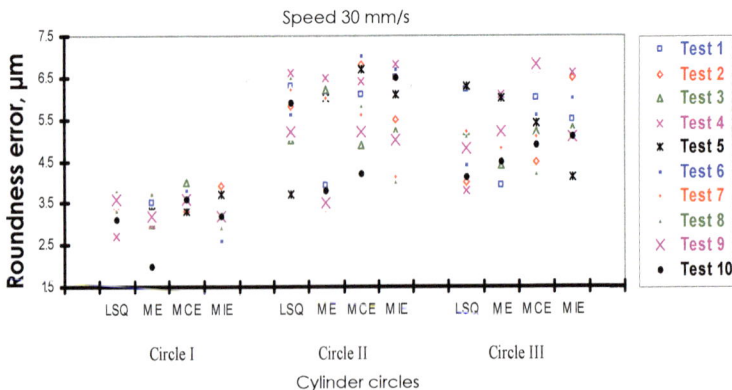

Figure 5.13   Roundness errors variation of fitting algorithms for three detection circles at 5 mm/s.

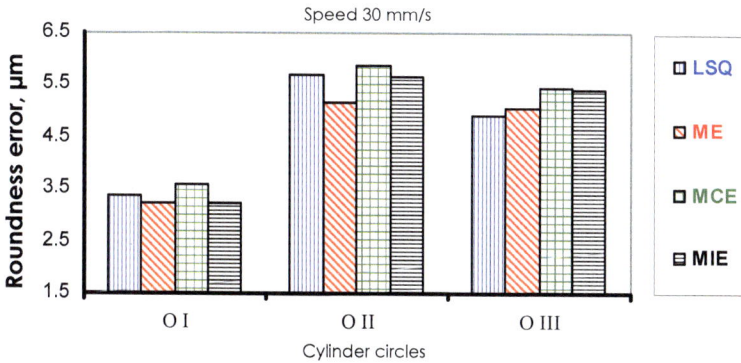

Figure 5.14 Measuring errors of fitting algorithms at 30 mm/s for three detection circles.

5.4.7   Probe Scanning 35 mm/s

Figure 5.15 shows the density of 120 measured points for roundness error using different four fitting algorithms of three transverse circles at probe scanning speed of 35 mm/s. Measured error results of roundness for fitting probe 35 mm/s are given in Fig. 5.16 for the detection circles. From the signal analysis, the following can be noticed:

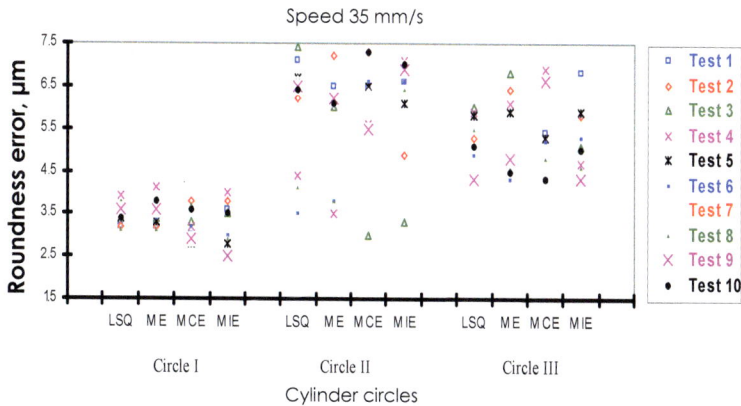

Figure 5.15 Measuring errors in the circles using different fitting algorithms.

- *Circle I* measurements have the lowest roundness error 3.25 µm for MCE and MIE fitting methods, the high error of 3.41 µm for the ME fitting method, and height error of 3.43 µm for the LSQ fitting method. Consequently, *circle I* has a measuring error range of 0.18 µm.

- The values of roundness measurement errors for all *circles II* and *III* have high difference of 0.75 and 0.53 µm, respectively.

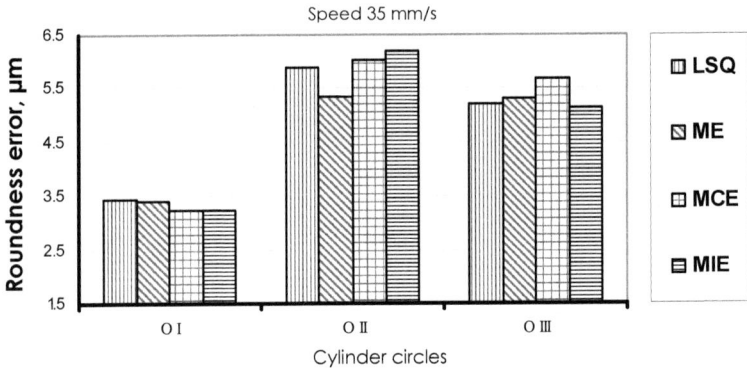

Figure 5.16 Roundness errors of fitting algorithms at 35 mm/s for three detection circles.

5.4.8 Probe Scanning Speed 40 mm/s

Figure 5.17 shows the density of measured points for roundness error using four fitting algorithms of three different transverse circles at probe scanning speed of 40 mm/s. The results in Fig. 5.18 show the measurement errors of roundness detection using different four fitting techniques. The analyses of the results illustrate the following:

- Roundness measurements of *circle I* have a minimum error of 3.1 µm for MIE fitting and a maximum error of 3.28 µm for ME response, while LSQ fitting error of 3.2 mm and MCE fitting error of 3.16 µm.
- Error measurements of *circle II* have a minimum of 5.28 µm for ME fitting and a maximum error of 5.99 µm for MCE response, while LSQ fitting error of 5.75 mm and MIE fitting error of 5.82 µm.
- Measurement for *circle III* have a minimum error of 4.69 µm for ME fitting and a maximum error of 5.06 µm for MIE response, while LSQ fitting error of 4.9 mm and MCE fitting error of 4.73 µm.

- Error difference for all three measured circles has a minimum value of 0.18 μm (3.28 and 3.1 μm) of *circle I* and a maximum value for the same measurements of 0.71 μm (5.99 and 5.28 μm) of *circle II*.

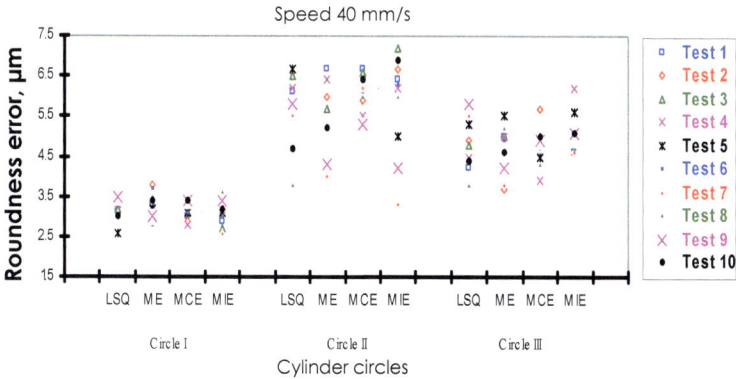

Figure 5.17   Roundness errors of fitting algorithms in the three circles using at 40 mm/s.

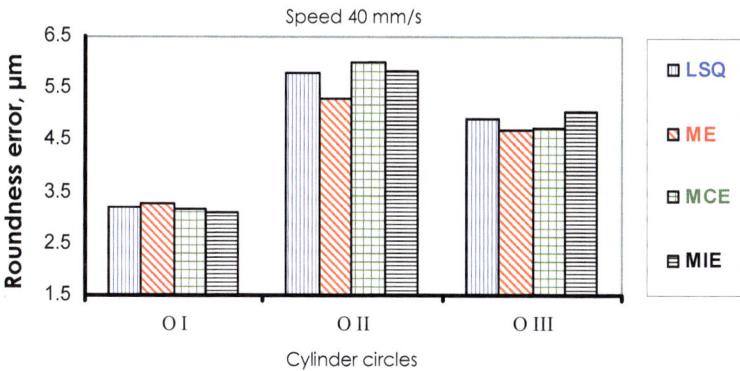

Figure 5.18   Measuring errors in the circles using different fitting algorithms.

5.4.9   Probe Scanning Speed 45 mm/s

Figure 5.19 shows the density of measured results for roundness error using different four fitting algorithms of three transverse circles at scanning speed of 45 mm/s. The results given in Fig. 5.20 show the measurement errors of roundness detection at probe scanning speed of 45 mm/s. The analyses of the results illustrate the following:

- Measurement of *circle I* have a minimum error of 3.25 µm for MIE fitting and a maximum error of 3.40 µm for ME response, while LSQ fitting error of 3.27 mm and MCE fitting error of 3.35 µm.
- Error measurements of *circle II* have a minimum of 4.88 µm for LSQ fitting and a maximum error of 5.38 µm for ME response, while MCE fitting error of 5.11 mm and MIE fitting error of 4.93 µm.
- Roundness Measurement for *circle III* have a minimum error of 5.01 µm for MIE fitting and a maximum error of 5.40 µm for ME response, while LSQ fitting error of 5.09 mm and MCE fitting error of 5.03 µm.

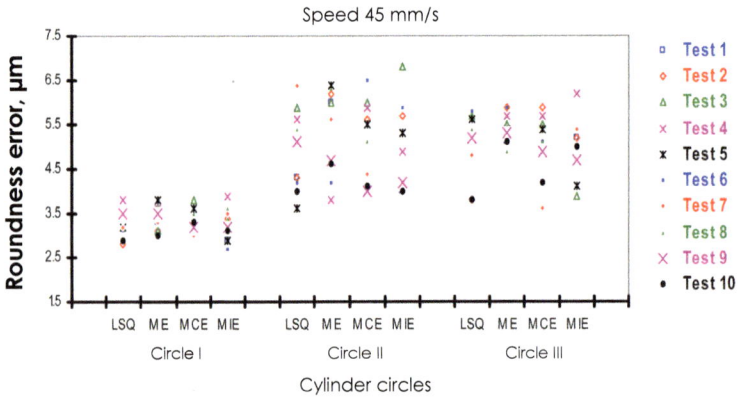

Figure 5.19 Roundness errors of fitting algorithms in the three circles using at 45 mm/s.

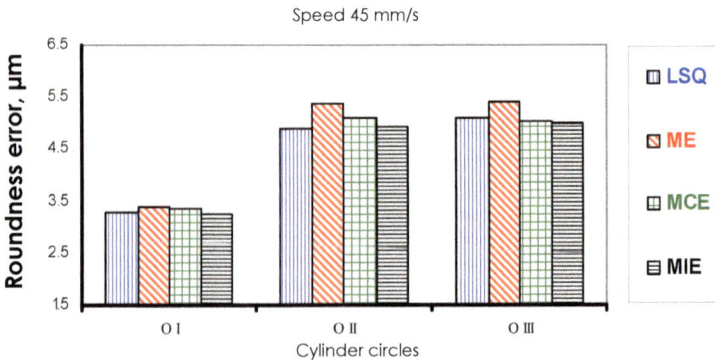

Figure 5.20 Measuring errors in the circles using different fitting algorithms.

5.5 Statistical Analysis

To make more reliable analysis of the influence of the fitting algorithm and the probe scanning speed on the CMM measurement accuracy, statistical tests should be used. The statistical analysis of the roundness error average and standard deviation mean value was calculated for selected parameters. The statistical results obtained are reduced and presented in a more practical and explicit error form in Figs. 5.21 and 5.22. The roundness errors result as function of the nine probe scanning speed using four fitting techniques given as follows.

5.5.1 Standard Deviation Average of Roundness Measurement Error

The averaging of the standard deviation error of the three circle signals is related to the fitting algorithm for the probe scanning speeds given in Table 5.2 and presented in Fig. 5.21. The results indicate the following:

Table 5.2 Standard deviation variation related to the probe speeds at different fitting algorithms

Probe speed (mm/s)	Standard deviation mean values (µm)				
	LSQ	ME	MCE	MIE	SD_{Mean}
5	0.3000	0.4700	0.8500	1.0600	0.6700
10	0.3000	0.3583	0.8167	1.0000	0.6188
15	0.3200	1.3334	1.1434	2.2533	1.2625
20	0.3500	1.3333	1.4800	2.0300	1.2983
25	0.4000	1.1867	1.3634	2.1567	1.2767
30	0.4000	1.1934	1.3967	2.0567	1.2617
35	0.4000	1.2634	1.3734	2.1834	1.3051
40	0.4000	1.1967	1.3367	2.0300	1.2409
45	0.4000	1.2800	1.4000	1.8534	1.2334
SD_{Mean}	0.3633	1.0684	1.2400	1.8471	$SD_{Average}$ 1.1297

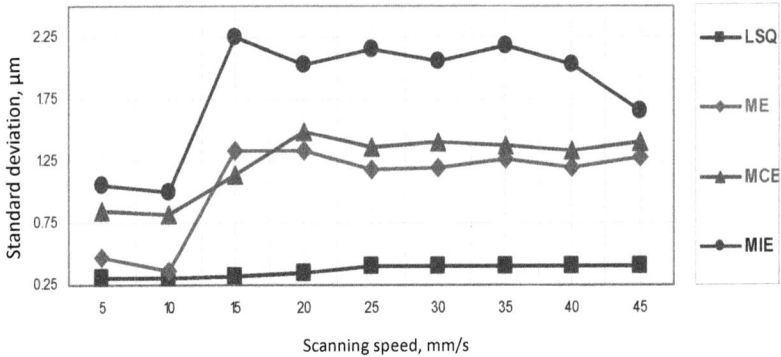

Figure 5.21 Standard deviations mean values of fitting algorithms at different probe speeds.

- Signal measured for evaluated test samples has a global average of 1.13 µm. The samples at probe scanning speeds from 5 to 45 mm/s (with 5 mm/s interval value) have the averaged values 0.62 to 1.31 µm, which correspond to represent 59.3%, 54.8%, 111.8%, 115%, 113%, 111.7%, 115.5%, 109.8%, and 109.2% of the global average, respectively. It ensures that measurement at probe speed before 15 mm/s is the suitable case for this work piece to satisfy the high level of accuracy.
- The roundness standard deviation has significant variation at 15 mm/s for all fitting methods; this may be due to probe response at resonance traveling speed.
- The MIE algorithm has highest standard deviation average response, while the LSQ algorithm has accurate response within the application range.

5.5.2 Roundness Error of Scanning Speed Response

The averaging of the roundness measured error of 50 mm ring circle signals of the fitting algorithms for probe scanning speed is given in Table 5.3 and presented in Fig. 5.22. The illustrated values indicate the following:

- Signal measured for 1080 evaluated samples has a global average of 4.14 µm. The samples at probe scanning speeds of 5, 10, 15, 20, 25, 30, 35, 40 and 45 mm/s have the averaged values 2.21, 1.87, 4.83, 4.95, 4.72, 4.72, 4.84, 4.59 and 4.51 μm, which correspond to represent 53.4%, 45.1%, 116.7%, 119.7%, 114.2%, 113.9%, 117.1%, 110.9% and 108.97% of the global average, respectively. It ensures that measurement at probe speed 10 and 5 mm/s are the suitable case for this work piece to satisfy the high level of accuracy.

Table 5.3 Roundness errors mean values related to the evaluation fitting algorithm at different probe speeds

Probe speed (mm/s)	Roundness error mean values (μm)				
	LSQ	ME	MCE	MIE	RON$_{Mean}$
5	2.2867	2.0533	2.2800	2.2200	2.2100
10	1.9333	1.7000	1.8133	2.0200	1.8667
15	4.8933	4.7867	4.8433	4.7800	4.8258
20	5.0100	4.8267	5.1233	4.8500	4.9525
25	4.7100	4.6033	4.6900	4.8900	4.7233
30	4.6500	4.4767	4.9567	4.7633	4.7117
35	4.8433	4.6867	4.9800	4.8667	4.8442
40	4.6367	4.4167	4.6267	4.6600	4.5850
45	4.4133	4.7267	4.4967	4.3933	4.5075
					RON$_{Average}$
RON$_{Mean}$	4.1530	4.0308	4.2011	4.1604	4.1363

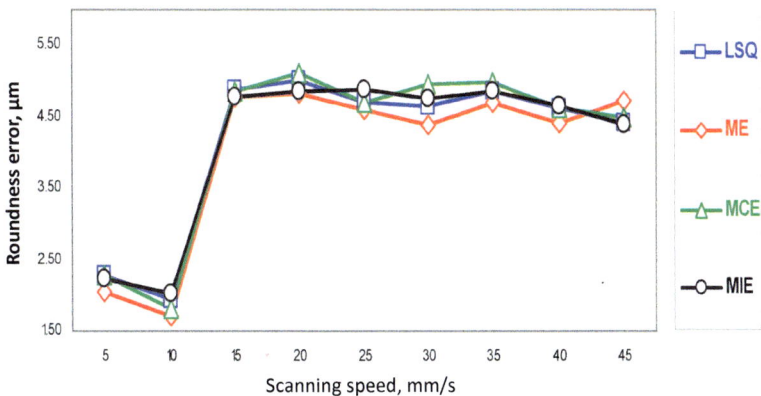

Figure 5.22 Influence of probe scanning speed on the roundness error for different fitting techniques.

- Averaged percentage errors as a function of the fitting algorithm response with respect to the global mean value are 100.4%, 97.5%, 101.6%, and 100.6%, which are corresponding to LSQ, ME, MCE, and MIE, respectively. Quality of measurements indicates that MCE and MIE methods have about 1.6% and 0.6% inaccuracy, while the LSQ algorithm has accurate responses with the error range of 0.4%.
- The roundness measurement error range has sharp significant variation at 15 mm/s compared to 5 and 10 mm/s scanning speeds.
- From the data presented in Fig. 5.22, the values have been treated statistically using polynomial regression type to get general formulae of the roundness error (RON) in μm as a function of probe scanning speed S for the different four fitting algorithms as follows:

$$RON_{(LSQ)} = 0.006S^4 - 0.1079S^3 + 0.5057S^2 + 0.1484S + 1.4213 \tag{5.1}$$

$$RON_{(ME)} = 0.0081S^4 - 0.1429S^3 + 0.6887S^2 - 0.175S + 1.3615 \tag{5.2}$$

$$RON_{(MCE)} = 0.0078S^4 - 0.1464S^3 + 0.7823S^2 - 0.556S + 1.8827 \tag{5.3}$$

$$RON_{(MIE)} = 0.006S^4 - 0.1111S^3 + 0.5452S^2 + 0.0196S + 1.4927 \tag{5.4}$$

From the above empirical Eqs. (5.1–5.4), the formulae illustrate that the MCE and MIE methods have higher error potentials of 1.9 and 1.5 μm to the probe scanning speed, respectively, where ME technique has a lowest error potential of 1.36 μm at high sensitivity coefficients of 0.18 to the probe scanning speed (S). The LSQ method has an error potential of 1.4 μm at sensitivity coefficients of 0.15 to the probe scanning speed and sensitivity coefficients of 0.5 μm to the probe scanning acceleration.

5.6 Conclusions

This chapter presents a new experimental investigation to improve the roundness measurement accuracy of a coordinate-measuring machine (CMM) for particular measurement tasks.

The method proposed requires just two selections of probe scanning speed and fitting algorithm for measuring the roundness of circles.

From this study to improve the roundness measurement quality, some conclusions can be drawn:

- There are roundness differences in the same work piece detecting circles; this may be due to selection of different fitting algorithms which have difference responses according to their software design within the maximum permissible scanning probing error.

- Suitable scanning speed of the machine touch probe should be well selected in accordance with the fitting algorithm to get high response quality with accurate roundness measurement.

- For similar roundness measurements, ME algorithm show high quality response beside high sensitivity coefficients to the probe scanning speed; both treatment methods ensure high measuring accuracy at low probe scan speed may be due to probe design resonance.

- Mean average of roundness error may be the reliable tool for CMM accuracy evaluation compared to standard deviation average within the application range.

- The range of roundness measuring error has the high significant value at 15 mm/s compared to 5 and 10 mm/s testing speeds for all fitting techniques. Result in the selected speed obtained significant variations, which may be due to probe response at resonance traveling speed.

Finally, the most basic measuring applications, the measurement of supplemental standard ring will increase our knowledge about the state of the measuring strategy.

Acknowledgment

The author appreciates the contributions of Polish Prof. Dr. (an *anonymous Polish reviewer*) who provided several comments and suggestions for improvement.

References

1. S. C. N. Topfer, G. Linss, M. Rosenberger, U. Nehse, and K. Weissensee, Automatic Execution of Inspection Plans with the I++DME

Interface for Industrial Coordinate Measurements, *M&MS Journal*, vol. XIV, no. 1 pp. 71–88, Poland, 2007.

2. C. J. Li and S. Y. Li, On-Line Roundness Error Compensation Via P-Integrator *Learning Control*, Columbia University, New York, NY, 10027, USA, 1991.

3. C. Diaz and Th. H. Hopp, Testing of Coordinate Measuring System Software, Proceedings of the 1993 American Society for Quality Control Measurement Quality Conference, USA, 1993.

4. Z. H. Xiong and Z. X. Li, *Error Compensation of Work Piece Localization*, RGC Grant No. HKUST 6220/98E and CRC98/0.1EG02, pp. 1–6, China, 2000.

5. P. Swornowski and M. Rucki, The Errors Occurring in the CMM Fitting Method, *Measurement Science Review*, vol. 3, Sec. 3, pp. 135–138, 2003.

6. A. Farooqui Sami and P. Morse Edward, Methods and Artifacts for Comparison of Scanning CMM Performance, *ASME*, vol. 7, issue 1, pp. 72–80, USA, 2007.

7. H. Hopp and M. Levenson, Performance Measures for Geometric Fitting in the NIST Algorithm Testing and Evaluation Program for Coordinate Measurement Systems, *Journal of Research of NIST*, vol. 100, issue 563, pp. 563–574, USA, 1995.

8. R. Fruhwirth, A. Strandlie, W. Waltenberger, and J. Wroldsen, A Review of Fast Circle and Helix Fitting, *Nuclear Instrument and Methods in Physics Research*, vol. 502, no. 2, pp. 705–707, 2003.

9. M. Abbe and K. Takamasu, *Modeling of Spatial Constraint in CMM Error for Uncertainty Estimation*, Proceedings of the 3rd Euspen International Conference, Eindhoven, Netherlands, May 26–30, 2002.

10. A. Forbes, Surface Fitting Taking into Account Uncertainty Structure in Coordinate Data, *Measurement Science and Technology*, vol. 17, pp. 553–558, 2006.

11. Edward P. Morse: *Artifact Section and Its Role in CMM Evaluation*, IDW, 2002.

12. M. Abbe and K. Takamasu, *Uncertainty Evaluation of CMM by Modeling with Spatial Constraint*, Proceedings of the 9th International Symposium on Measurement and Quality Control (9th ISMQC), pp. 121–125, IIT Madras, India, November 21–24, 2007.

13. International Standard: *Geometrical Product Specifications (GPS)— Acceptance and Reverification Tests for Coordinate Measuring*

Machines (CMM)—Part 2: CMMs used for Measuring Size, ISO 10360-2, 2nd edition, 15-12-2001.

14. International Standard: *Geometrical Product Specifications (GPS)— Acceptance and Reverification Tests for Coordinate Measuring Machines (CMM)—Part 4: CMMs used in Scanning Measuring Mode*, ISO 10360-4, 2nd edition, 15-3-2000.

Chapter 6

Validation Method for CMM Measurement Quality Using Flick Standard

6.1 Introduction

Micro coordinate metrology is an important branch of product quality assurance. Coordinate measuring machine (CMM) is generally better optimized in micro coordinate metrology. CMM is an advanced computerizing numerical control system that can be used for dimensional inspection of complex 3D geometry products. The advanced CMM is one of the most diversified fields of mechanical engineering due to the promising technology cooperation with modern industrial strategies. Surface quality is an important parameter in the modern manufacturing, aerospace, and automotive. The influence of dimensional and geometrical form errors is one of the issues of interest regarding manufacturing accuracy and surfaces quality. Out of tolerance can produce number of problems like vibration, wear, and noise [1, 2]. The study of the influence of the geometric fitting algorithms and probe scanning speeds on the measurement error can be an interesting topic for future research, especially with the growing use of CMM machines [2–5]. In addition, the development of a new CMM machine with accuracy less than a

Automotive Engine Metrology
Salah H. R. Ali
Copyright © 2017 Pan Stanford Publishing Pte. Ltd.
ISBN 978-981-4669-52-8 (Hardcover), 978-1-315-36484-1 (eBook)
www.panstanford.com

nanometer is necessary, because it is close to the actual industrial needs. For these reasons, most available validation and calibration of CMM require reference standards artifacts. Flick standard artifact is embodiment of outer surface. New published work in parallel with my research, confirmed that the Flick standard artifact is a suitable tool used to calibrate the spindle motion error and the probe which equips cylindrical measuring machines to ensure such a level of uncertainty, both stability and performance [5]. The resultant measurement quality of CMM is limited by deviation and some uncertainties due to different factors. The measurement deviations in CMM metrology can be related to the operator performance quality, environmental interaction, and CMM performance accuracy. It can be assumed that some important factors such as operator behavior and CMM software accuracy have effective reactions on the measurement quality. Uncertainty in measurement is the most important concept in roundness metrology according to the GUM and attracts wide attention in the world to confirm the confidence in measurement result [6].

In this chapter, *measurements* for the transverse circle location of certified Flick standard artifact *have been* carried out. The measurement results of dimension and geometrical form at different CMM strategies have been studied. The influence of geometric fitting algorithms equipped with probe scanning speeds as metrological parameters on both diameter (D) and roundness form errors (RON) is verified experimentally and discussed. The modifications and analysis of software are employed experimentally. The statistical *study* based on design of *experiment* was accomplished in order to *evaluate* the measurement result and estimate the expanded uncertainties experimentally and statistically. Therefore, the main objective of this research is to study the influence of four fitting algorithms through five different probe scanning speeds for transverse circle location of carrying out signals, to

(a) develop the CMM software using closed loop control to develop new accurate and precise machines;

(b) increase CMM operator skills, reduce the operation lost time and cost, to avoid processing mistakes of software strategic applications;

(c) minimize the size of uncertainty in measurement;

(d) determine and analyze the relative deviation errors in measurement.

6.2   Experimental Work

The experiment presented in this chapter is designed and performed within the research plan of the engineering and surface metrology department at NIS. *The* new validation method including measurement procedures is designed to *study* the effect of geometric fitting algorithm and probe scanning speed in CMM measurement. This plan has been done in two main stages. The first stage is the verification of the CMM probing system. The main stage in the experiments is made to verify the CMM machine in measurement quality. This second stage based on measure the diameter and roundness of the Flick standard artifact at different strategies in measurement. The dynamic influences of geometric fitting algorithms and probe canning speeds strategies on diameter and roundness deviations have been studied using CMM.

6.2.1   Dynamic Verification of Probing System

Dynamic verification of probing system is a very important recommendation task for an acceptance before studying CMM performance accuracy [7, 8]. The standard verification method of both probing error and scanning probing error using a reference sphere is used. To determine the probing error, 25 recommended points on the reference test sphere surface must be probed. To determine the scanning probing error, four recommended scanning lines on the surface of test sphere must be scanned and the Gaussian center point of the sphere using all measured points of all four scan lines must be computed. Before making measurements with the CMM in the cylindrical feature of Flick standard artifact, the CMM was verified using master probe for evaluate standard sphere and using standard sphere for evaluate used probe [9]. The standard deviation (SD) of program output result in the CMM verification test conditions is presented in Table 6.1.

Table 6.1   Output verification data of CMM probes and sphere

CMM element	Measured value (mm)	SD (mm)
Master probe	$R = 3.9999$	0.0002
Reference sphere	$R = 14.9942$	0.0002
Used stylus probe	$R = 1.4990, L = 33.5$	0.0002
With Al. extension	$L = 100$	
Step width	1.3861	
No. of probe points	104 points	

The CMM has limited specific values as follows:

$$\text{MPE} = A + L/K, \mu m, \text{MPE}_P = 1.00 \ \mu m; \text{MPET}_{ij} = 1.90 \ \mu m,$$

where MPE is the maximum permissible measurement error, A is the constant machine uncertainty equal to 0.9 μm, K is the length constant or slope of line equal to 350, and L is the length measurement in mm. MPE_P is the maximum permissible probing error and MPET_{ij} is the maximum permissible error when measuring a part by using scanning mode which is called maximum permissible scanning probing error.

6.2.2   Flick Standard Artifact

The best method of dynamic validation is made through the use of a high-precision Flick standard in roundness Talyrond machines. Existing outer surface of Flick standard artifact is only one applicable embodiment type of sensitivity validation for dimensional and geometrical form. Flick standard artifact is a cylindrical artifact with a known nominally one flat face through the outer perimeter. Flick standard artifact has some major disadvantages like a small dynamical contact for larger wave numbers [10]. The amplitude of signals for a typical Flick with a cylinder diameter of about 44 mm and a deepness of the flattening of (294.9:283.3) μm have been measured (see Fig. 6.1). While the measured signal to noise ratio is actually low for reference standard Flicks. Therefore, the author predicted that the Flick standard artifact might be a new suitable tool for validation method of CMM measurement quality.

The CMM output significant contribution from larger wave numbers than 25 UPR (undulation per revolution) is demonstrated in Fig. 6.2. The output result has been performed in an appropriate measurement range for the diameter and roundness form deviations. But the calibration result depends on the individual wave amplitudes. The output signal to noise ratio is quite low for Flick, because the full measured signal is carried by a single wave number.

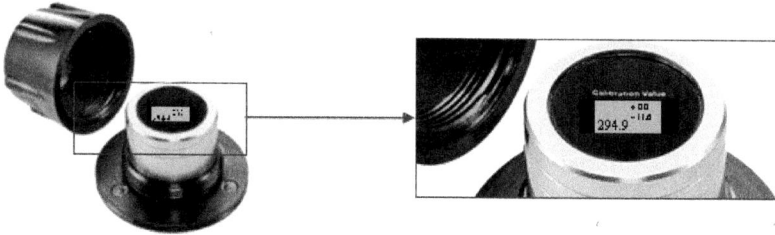

Figure 6.1   Flick standard artifact.

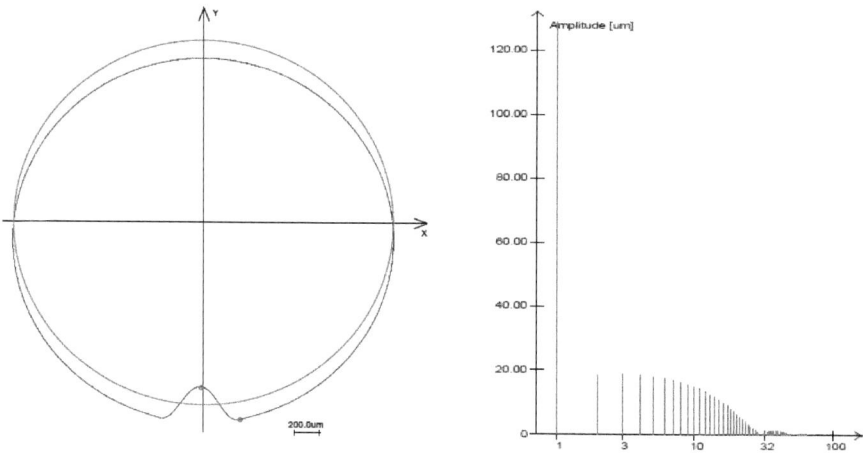

Figure 6.2   Typical result of a Flick standard artifact at scanning speed of 1 mm/sec at LSQ fitting algorithm and Gaussian filter using CMM with magnification factor 100. (a) Outer roundness form profile in time domain. (b) Spectrum amplitude in frequency domain.

6.2.3   CMM Measurement Procedures

The measurements have been carried out at the same transverse section on the outer surface of the Flick standard artifact at location of 18 mm from the top to detect any errors of the surface. The measured sample errors were obtained for five probe scanning speeds 1, 2, 3, 4, and 5 mm/s during 360° angle range trace of the Flick standard, respectively. While CMM traveling speed was constant at 70 mm/s and the number of scan fitting points also was constant with about 104±2 points during measurement tests at temperature condition of 20±0.5°C. Each measurement point has 10 times repetitions for the same transverse circle (x, y and z) positions. The feature of roundness measurement has been determined at each probe scanning speed, where the CMM PRISMO navigator has been selected corresponding to evaluation fitting algorithms. Four main geometric fitting algorithms are recommended in CMM measurement for cylindrical parts (LSQ, ME, MC, and MI), where LSQ means least square fitting algorithm, ME means minimum element fitting algorithm, MC is minimum circumscribed fitting algorithm, and MI means maximum inscribed fitting algorithm. Determination of measured deviation errors has been included 200 experimental measuring tests to differentiate between evaluation qualities of different measurement strategies. The diameter and roundness deviation errors of corresponding evaluation algorithm at different scanning speeds are obtained through comparison, judgment, and repeated arrangement. However, a question arises: Which algorithm is suitable to choose and what criterion should be taken at which probe scanning speed?

6.3 Measurement Results and Discussion

The distribution density of measured points of both diameter and roundness variation is presented in Figs. 6.3, 6.5, 6.7, 6.9, 6.11, 6.13, 6.15, and 6.17. The errors average of results obtained for both diameter and roundness variation are reduced and presented in a more practical and explicit form in Figs. 6.4, 6.6, 6.8, 6.10, 6.12, 6.14, 6.16, and 6.18. The diameter and roundness error results as functions of the probe scanning speed and fitting algorithm are given as follows.

6.3.1 Least Square Fitting Technique

The results in Fig. 6.3 present the distribution density of measured points for diameter error of 10 reputed test results for Flick artifact transverse circle using LSQ fitting algorithm at different probe scanning speeds. Figure 6.4 shows the average variation of diameter error for different probe scanning speeds using LSQ fitting algorithm. The analysis of the 50 given results indicates the following:

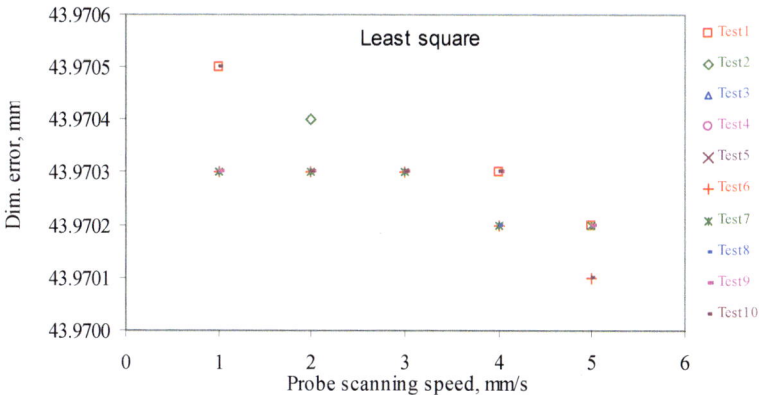

Figure 6.3 Diameter errors variation of LSQ fitting algorithm at different probe scanning speeds.

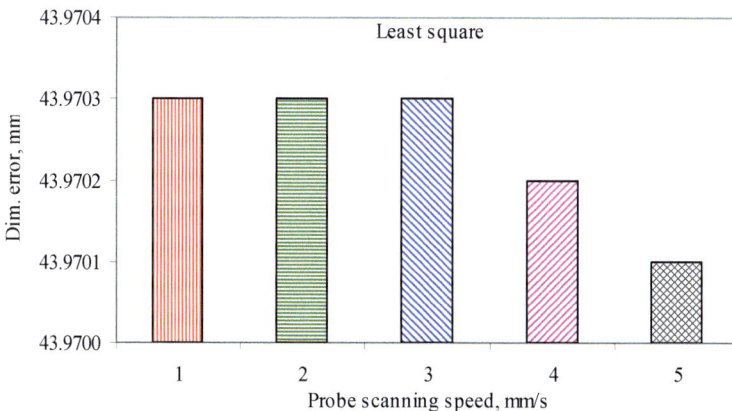

Figure 6.4 Measuring errors average of diameter at different probe scanning speeds with LSQ fitting algorithm.

- Detection LSQ fitting algorithm at probe speed of 1mm/s measurements has a diameter error range of 0.2 μm from 43.9705 mm to 43.9703 mm, while measurements at 2 mm/s have diameter error limits of 43.9704 and 43.9703 mm with an error range of 0.1 μm, while the error ranges at 3, 4, and 5 mm/s measurements became stable within the same value of 0.1 μm.

- The application based on the LSQ fitting technique in all measuring speeds clearly showed that, the evaluated difference of average error between probe speeds as representing values to the scanning speed quality has 0.1 and 0.2 μm for 1, 2, and 3 mm/s (43.9703 mm) and 4 mm/s (43.9702 mm), while 5 mm/s (43.9701 mm), respectively.

- Measurements at 4 mm/s have the low diameter error for the LSQ fitting algorithm, while at 5 mm/s have the lowest diameter error.

The results in Fig. 6.5 show the distribution density of measured points for roundness error of 10 reputed test results for Flick transverse circle using LSQ fitting algorithm at different probe scanning speeds. Figure 6.6 shows the average variation of roundness error for different probe scanning speeds using LSQ fitting algorithm. The result analysis of the 50 tests indicates the following:

- Roundness measurements of LSQ algorithm at 2 mm/s have a minimum error of 0.2 μm and maximum error of 0.5 μm at 5 mm/s.

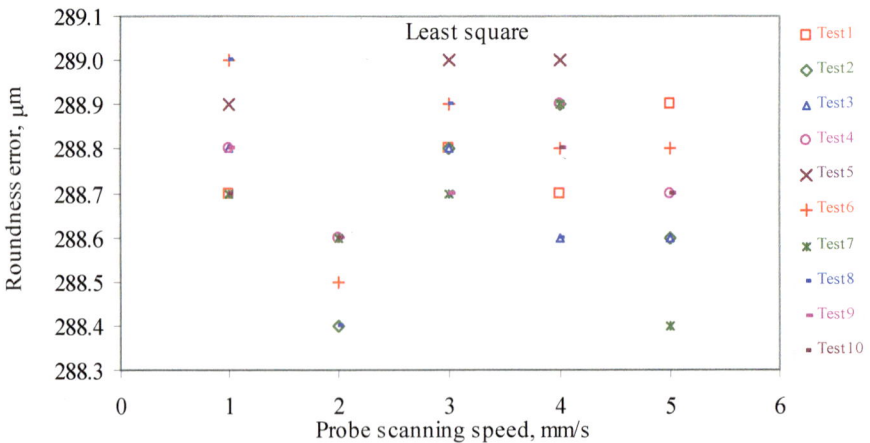

Figure 6.5 Roundness errors variation of LSQ fitting algorithm at different probe scanning speeds.

- The application of the LSQ fitting technique to all measuring speeds showed that the evaluated difference of average error between probe speeds as representing values to the scanning speed quality has 0.3 for 1 and 4 mm/s (288.8 μm) and 2 mm/s (288.5 μm), while at 3 mm/s (288.9 μm) and (288.7 μm) at 5 mm/s.

- The value of roundness measurement errors for 2 mm/s has the lowest significant variation compared to all other probe speeds.

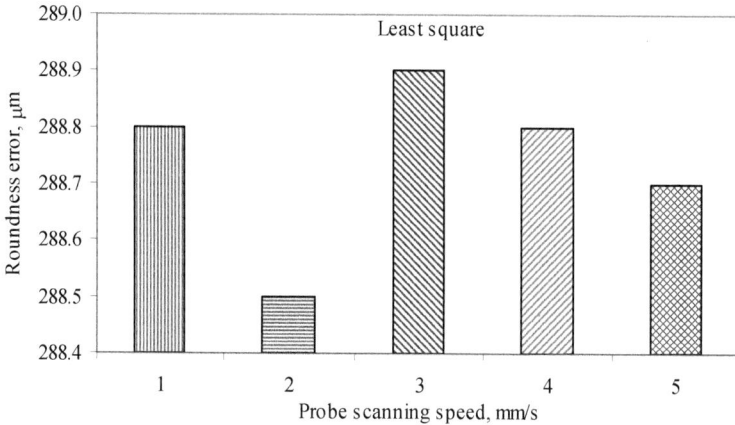

Figure 6.6 Measuring errors average of roundness at different probe scanning speeds with LSQ fitting algorithm.

6.3.2 Minimum Element Fitting Technique

Figure 6.7 shows the density of measured points for diameter error of 10 reputed test results for Flick artifact transverse circle using ME fitting algorithm at different probe scanning speeds. Figure 6.8 shows the average variation of the diameter error for different probe scanning speeds using ME fitting algorithm. The analysis of the 50 given results indicates the following:

- Detection ME fitting algorithm at probe speed of 1 mm/s measurements has a diameter error range of 0.5 μm from 43.9868 mm to 43.9863 mm, while measurements at 2 mm/s have diameter error limits of 43.9865 and 43.9864 mm with an error range of 0.1 μm. The error ranges at

3 mm/s measurements have the highest diameter error range of 0.6 μm. Measurements at 2 mm/s have the lowest diameter error for the ME fitting algorithm, while at 1 mm/s have the high diameter error of 0.5 μm.

- According to the application of the ME fitting algorithm to all measuring speeds, the evaluated difference of average error between probe speeds as representing values to the scanning speed quality has 0.1 and 0.6 μm for 1, 3, and 4 mm/s (43.9865 mm), while at 2 and 5 mm/s (43. 9864 mm), respectively.

- The measuring error range has the lowest significant variation at 2 mm/s compared to 1, 3, 4, and 5 mm/s testing speed, which may be due to probe response at resonance traveling speed.

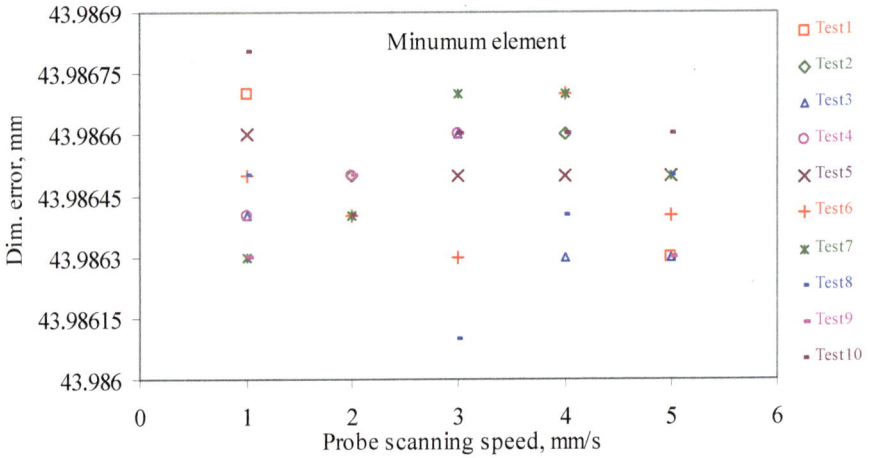

Figure 6.7   Diameter errors variation of ME fitting algorithm at different probe scanning speeds.

The results presented in Fig. 6.9 show the distribution density of measured points for roundness error of 10 reputed test results for Flick transverse circle using ME fitting algorithm at different probe scanning speeds. Figure 6.10 shows the average variation of roundness error for different probe scanning speeds using ME fitting algorithm. The analysis of the given result indicates the following:

- Roundness measurements of ME algorithm at 2 mm/s have a minimum average error of 285.7 μm relative to all other probe speeds.
- According to the application of the ME fitting technique to all measuring speeds, the evaluated difference of average error between probe speeds as representing values to the scanning speed quality is 286.0 μm for 1, 3, 4 and 5 mm/s.
- The value of roundness measurement errors for 2 mm/s has the lowest significant variation of 0.3 μm compared to all other probe scanning speeds.

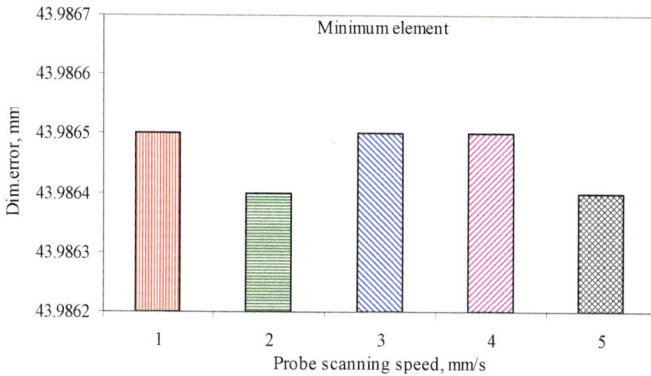

Figure 6.8   Measuring errors average of diameter at different probe scanning speeds with ME fitting algorithm.

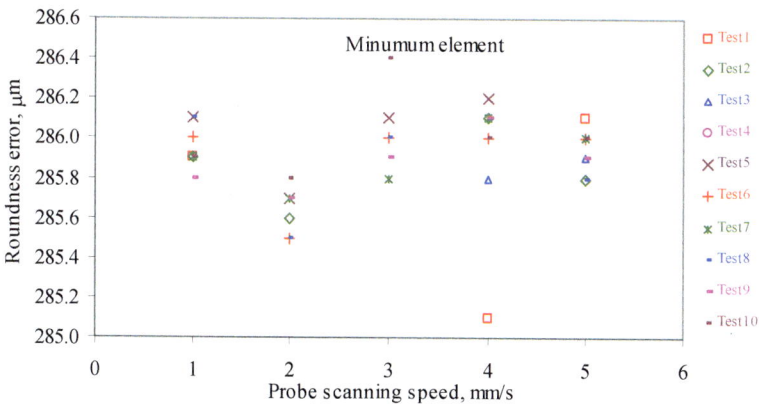

Figure 6.9   Roundness errors variation of ME fitting algorithm at different probe scanning speeds.

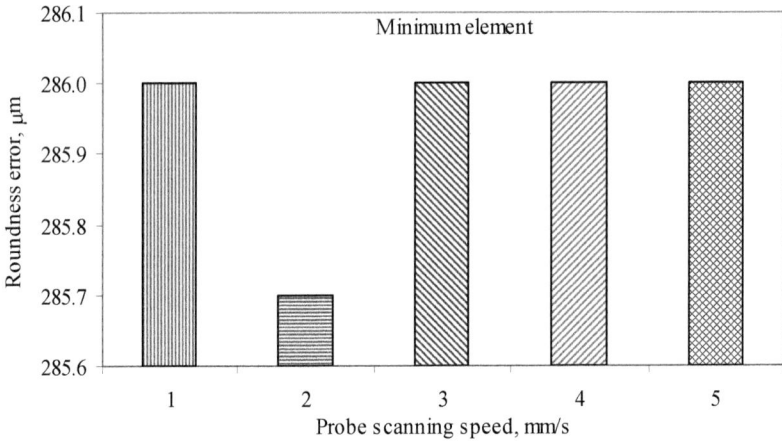

Figure 6.10 Measuring errors average of roundness at different probe scanning speeds with ME fitting algorithm.

6.3.3   Minimum Circumscribed Fitting Technique

The results in Fig. 6.11 show the density of measured points for diameter error of 10 reputed test results for Flick artifact transverse circle using MC fitting algorithm at different probe scanning speeds. Figure 6.12 shows the average variation of diameter error for different probe scanning speeds using MC fitting algorithm. The analysis of the 50 given results indicates the following:

- Detection MC fitting technique at probe speed of 1 mm/s measurements have a diameter error range of 0.3 µm from 43.9907 mm to 43.9904 mm, while measurements at 2 mm/s have diameter error limits of 43.9907 and 43.9905 mm with an error range of 0.2 µm. While the error ranges at 3 and 5 mm/s measurements have the lowest diameter variation error range of 0.1 µm. Measurements at 5 mm/s have the lowest average diameter error of 43.9903 mm for the ME fitting algorithm.
- According to the application of the MC fitting algorithm to all measuring speeds, the evaluated difference of average error between probe speeds as representing values to the scanning speed quality, has 0.1 and 0.3 µm for 1 and 2 mm/s (43.9906 mm), for 3 and 4 mm/s (43.9905 mm), while 5 mm/s (43.9903 mm), respectively.

- The measuring error range has the lowest significant variation at 3 mm/s compared to 1 mm/s testing speed, which may be due to probe response at resonance traveling speed at MC algorithm design.

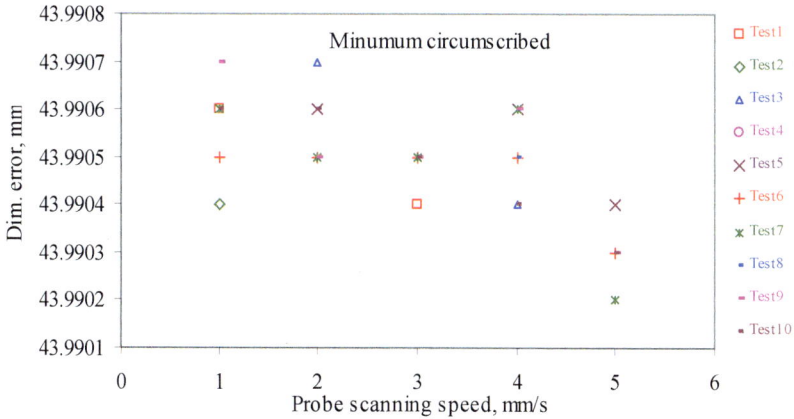

Figure 6.11   Diameter errors variation of MC fitting algorithm at different probe scanning speeds.

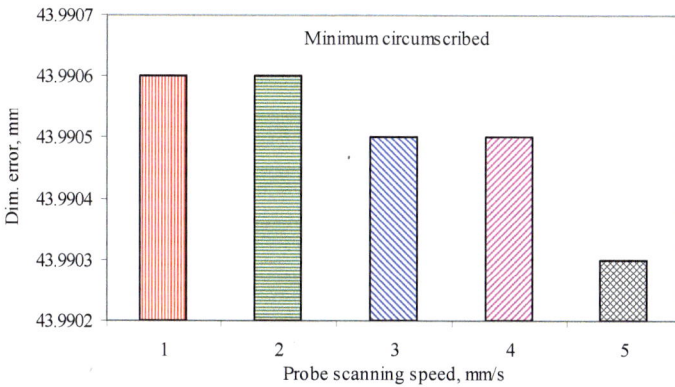

Figure 6.12   Measuring errors average of diameter at different probe scanning speeds with MC fitting algorithm.

The results in Fig. 6.13 show the density of measured points for roundness error of 10 reputed test results for Flick transverse circle using MC fitting algorithm at different probe scanning speeds. Figure 6.14 shows the average variation of roundness error for different probe scanning speeds using MC fitting technique. The result analysis of the 50 tests indicates the following:

- Roundness measurements of MC algorithm at 1, 3, and 4 mm/s have a minimum error of 0.4 μm and maximum error of 0.5 μm at 2 and 5 mm/s.
- According to the application of the MC fitting technique to all measuring speeds, the evaluated difference of average error between probe speeds as representing values to the scanning speed quality, has 0.4 μm for 1 and 4 mm/s (289.4 μm) and 2 mm/s (289.1 μm), while at 3 mm/s (289.5 μm) and (289.3 μm) at 5 mm/s.
- The value of roundness measurement errors for 2 mm/s has the lowest significant variation compared to all other probe scanning speeds.

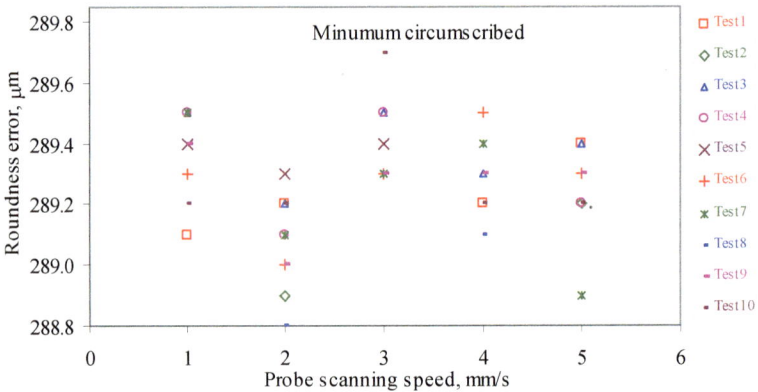

Figure 6.13 Roundness errors variation of MC fitting algorithm at different probe scanning speeds.

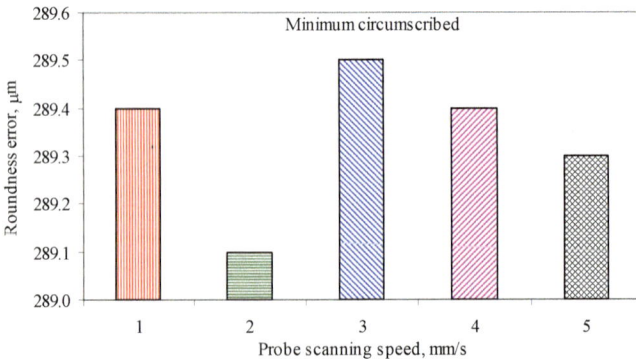

Figure 6.14 Measuring errors average of roundness at different probe scanning speeds with MC fitting algorithm.

6.3.4   Maximum Inscribed Fitting Technique

Figure 6.15 shows the density of measured points for diameter error of 10 reputed test results for Flick artifact transverse circle using MI fitting algorithm at different probe scanning speeds. Figure 6.16 shows the average variation of diameter error for different probe scanning speeds using MI fitting algorithm. The analysis of the 50 given results indicates the following:

- Detection MI fitting technique at probe speed of 1 mm/s measurements have a diameter error range of 0.5 μm from 43.7013 to 43.7008 mm, while measurements at 2 mm/s have diameter error limits of 43.7015 and 43.7011 mm with an error range of 0.4 μm. The error ranges at 3 and 5 mm/s measurements have the highest diameter error range of 0.5 μm. Measurements at 4 mm/s have the lowest diameter error variation of 0.2 μm for the MI fitting algorithm.

- According to the application of the MI fitting algorithm to all measuring speeds, the evaluated difference of average error between probe speeds as representing values to the scanning speed quality has 0.5 and 0.2 μm for 1 and 4 mm/s (43.7010 mm), for 3 and 5 mm/s (43.9709 mm), while 2 mm/s (43.9713 mm), respectively.

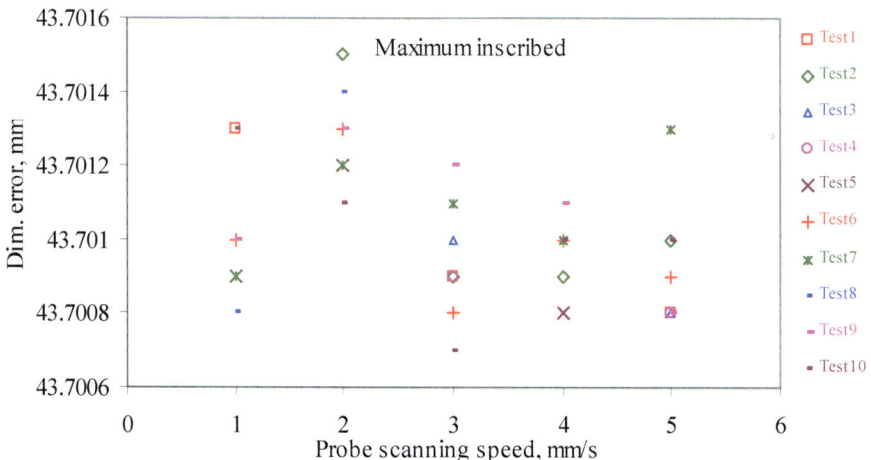

Figure 6.15   Diameter errors variation of MI fitting algorithm at different probe scanning speeds.

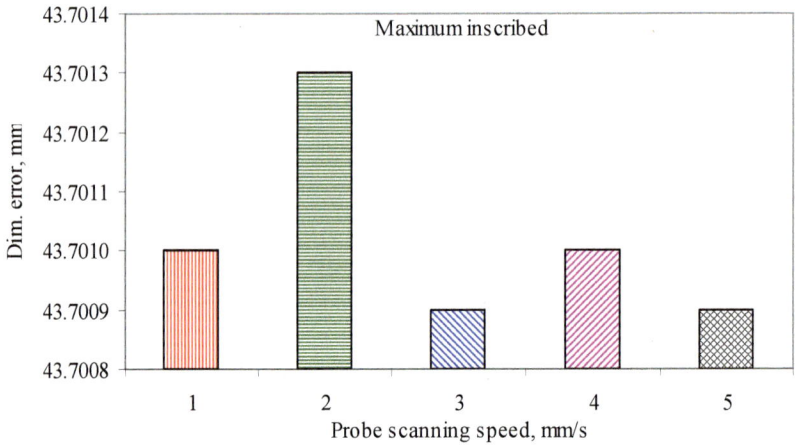

Figure 6.16 Measuring errors average of diameter at different probe scanning speeds with MI fitting algorithm.

The results presented in Fig. 6.17 show the distribution density of measured points for roundness error of 10 reputed test results for Flick transverse circle using MI fitting algorithm at different probe scanning speeds. Figure 6.18 shows the average variation of roundness error for different probe scanning speeds using MI fitting algorithm. The analysis of the 50 given results indicates that the following:

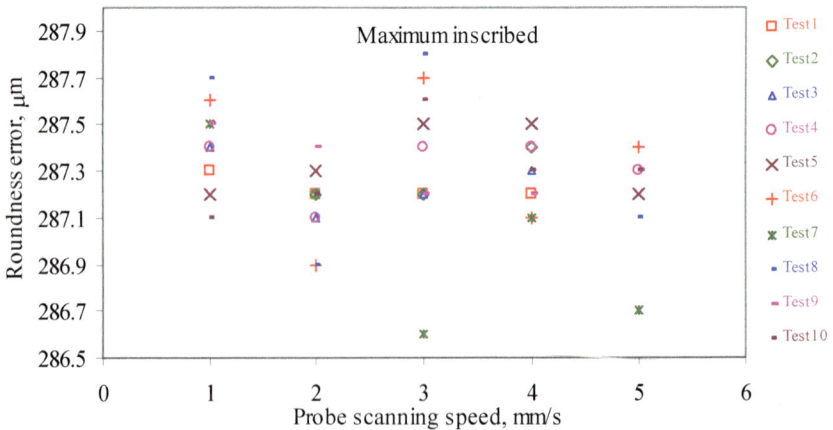

Figure 6.17 Roundness errors variation of MI fitting algorithm at different probe scanning speeds.

- Roundness measurements of MI algorithm at 4 mm/s have a minimum error of 0.4 μm and maximum error of 0.6 μm at 1, 3 and 5 mm/s.
- According to the application of the MI fitting technique to all measuring speeds, the evaluated difference of average error between probe speeds as representing values to the scanning speed quality, has 287.3 μm at 3 and 4 mm/s, while at 1 mm/s (287.4 μm) and (287.2 μm) at 2 and 5 mm/s.
- The value of roundness measurement errors for 2 and 5 mm/s has the lowest significant average variation compared to other probe speeds.

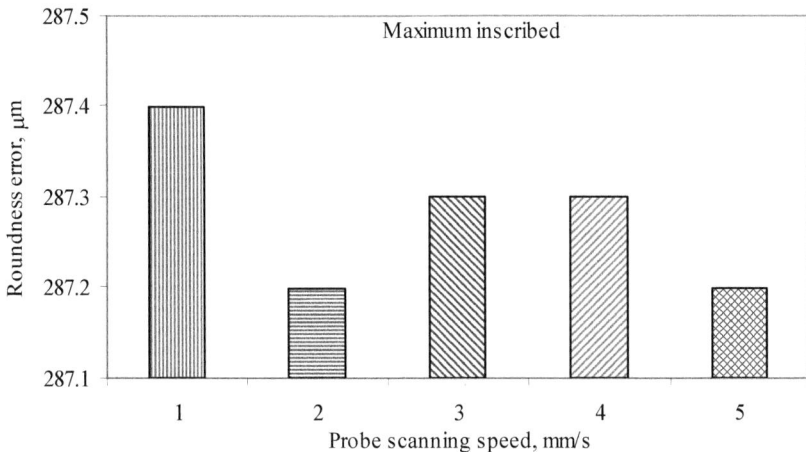

Figure 6.18   Measuring errors average of roundness at different probe scanning speeds with MI fitting algorithm.

6.4   Statistical Analysis

To give a more reliable analysis of the experimental data of the influence of fitting algorithm and probe scanning speed on the CMM measurement accuracy, statistical tests should be used. Along with the statistical analysis of diameter and roundness error averages, standard deviation and combined uncertainty (u_1) due to repeatability are calculated. Expanded uncertainty for selected parameters in measurement has been evaluated. The statistical results obtained are reduced and presented in a more practical and explicit error form in Figs. 6.19, 6.20. The errors averages in diameter and roundness results as function of the five probe scanning speed using four fitting techniques are given as follows.

6.4.1   The Error in Diameter Measurement

The error response of diameter variation for Flick standard artifact related to different CMM measurements strategies of different fitting algorithms of probe scanning speeds indicated in Table 6.2 and presented in Fig. 6.19. The indicated results include that:

- Signal measured for 200 evaluated samples has a global average of Flick diameter of 43.9120 mm. The samples at probe scanning speeds of 1, 2, 3, 4, and 5 mm/s have the roundness averaged values 43.9121, 43.9122, 43.9121, 43.9121, and 43.9119 mm, respectively, which correspond to represent 100.0002%, 100.0005%, 100.0002%, 100.0002%, and 99.9998% of the global average, respectively. It ensures that measurements at probe speed 5 mm/s are the suitable case for this work piece Flick to satisfy the high level of accuracy.

- Averaged percentage errors as a function of the fitting algorithm response with respect to the global mean value are 100.1325%, 100.1697%, 100.1788%, and 99.5195%, which correspond to LSQ, ME, MC, and MI, respectively. The quality of measurements indicates that MC and ME algorithms have about 0.1788% and 0.1697% inaccuracy, while the LSQ and MI algorithms have accurate responses with the error range of 0.1325% and −4805%, respectively.

- Signal measured for evaluated test samples at probe scanning speeds from 1 to 5 mm/s (with 1 mm/s interval value) have the standard deviation (SD) values of 0.1221, 0.1220, 0.1221, 0.1221, and 0.1221 mm, respectively. It ensures that measurement at probe speed of 2 mm/s is the suitable case for this work piece to satisfy the high level of accuracy.

- The diameter standard deviation has best significant variation at 2 mm/s for all fitting algorithms.

- The ME algorithm has lowest standard deviation average response, while all other algorithms have the same accurate response within the application range.
- The uncertainty due to repeatability in diameter measurement for four fitting algorithms does not exceed 0.0001.
- From the data presented in Fig. 6.19, the values have been treated statistically using fit a one degree linear regression type to get general formulae of the diameter error in mm as a function of probe scanning speed V for the different four fitting algorithms as follows:

Table 6.2 Average variation and standard deviation of Flick diameter (mm)

Scanning speed (m/s)	LSQ	ME	MC	MI	Average of D	SD
1	43.9703	43.9865	43.9906	43.7010	43.9121	0.1221
2	43.9703	43.9864	43.9906	43.7013	43.9122	0.1220
3	43.9703	43.9865	43.9905	43.7009	43.9121	0.1221
4	43.9702	43.9865	43.9905	43.7010	43.9121	0.1221
5	43.9701	43.9864	43.9903	43.7009	43.9119	0.1221
Average of D	43.9702	43.9865	43.9905	43.7010	43.9120	
SD	0.0001	0.0000	0.0001	0.0001		
$u_1 = SD/\sqrt{n}$	0.0001	0.0000	0.0001	0.0001		

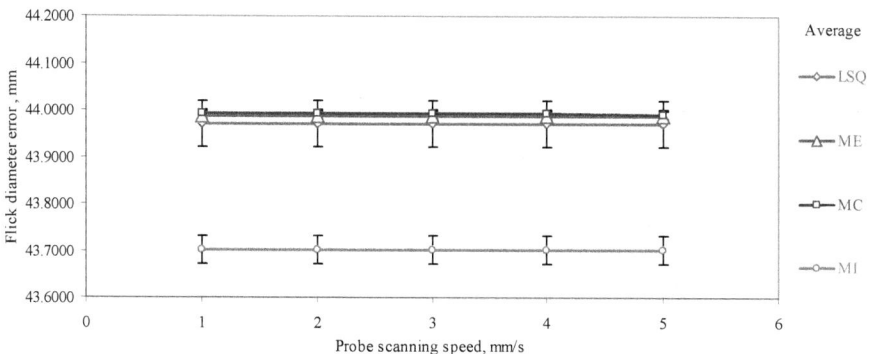

Figure 6.19 Average variation of Flick artifact diameter measurements at different CMM strategies.

$$RON_{(MC)} = -1E\text{-}05V + 43.9860, \tag{6.1}$$

$$RON_{(ME)} = -7E\text{-}05V + 43.9910, \tag{6.2}$$

$$RON_{(LSQ)} = -5E\text{-}05V + 43.9700; \tag{6.3}$$

$$RON_{(MI)} = -5E\text{-}05V + 43.7010 \tag{6.4}$$

From the linear regression Eqs. (6.1–6.4), the empirical formulae illustrate that the ME algorithm has higher an error potential of 43.991 mm at sensitivity coefficients of –7E-05 to the probe scanning speed (*V*), where MI technique has a lowest error potential of 43.701 mm at sensitivity coefficients of –5E-05 to the probe scanning speed. The LSQ algorithm has an error potential of 43.970 mm at sensitivity coefficients of –5E-05 to the probe scanning speed. The MC algorithm has an error potential of 43.986 mm at sensitivity coefficients of –1E-05 to the probe scanning speed.

6.4.2 The Error in Roundness Measurement

The error response of roundness variation for Flick standard artifact related to different CMM measurements strategies of different fitting algorithms of probe scanning speeds is presented in Table 6.3 and shown in Fig. 6.20. The statistical results indicate the following:

- Signal measured for 200 evaluated samples has a global average of Flick roundness of 287.8 μm. The samples at probe scanning speeds of 1, 2, 3, 4, and 5 mm/s have the roundness averaged values 287.9, 287.6, 287.9, 287.9, and 287.8 μm, respectively, which correspond to represent 100.04%, 99.93%, 100.04%, 100.04%, and 100.00% of the global average, respectively. It ensures that measurement at probe speed 2 mm/s are the suitable case for this work piece Flick artifact to satisfy the high level of accuracy.
- Averaged percentage errors as a function of the fitting algorithm response with respect to the global mean value are 100.31%, 99.48%, 100.73%, and 99.83%, which correspond to LSQ, ME, MC, and MI, respectively. The quality of measurements indicates that LSQ and MC algorithms

have about 0.31% and 0.73% inaccuracy, while the ME algorithm has accurate responses with the error range of 0.52%.

- Signal measured for evaluated test samples at probe scanning speeds from 1 to 5 mm/s (with 1 mm/s interval value) have the standard deviation (SD) values of 1.3153, 1.3065, 1.3718, 1.3255 and 1.2903 mm, respectively. It ensures that measurement at probe speed of 5 mm/s is the suitable case for this work piece to satisfy the high level of accuracy.

- The roundness standard deviation has significant variation at 3 mm/s for all fitting algorithms.

- The LSQ and MC algorithms have highest standard deviation average response, while the MI algorithm has accurate response within the application range.

- The uncertainty due to repeatability in roundness measurement for four fitting algorithms lies between 0.0606 and 0.0335.

- As shown in Fig. 6.20, the values have been treated statistically using fourth-order polynomial regression fit type to get general formulae of the roundness error (RON) in μm as a function of probe scanning speed V for the different four fitting algorithms as follows:

Table 6.3 Average variation and standard deviation of Flick roundness (μm)

Scanning speed (mm/s)	LSQ	ME	MC	MI	Average of RON	SD
1	288.8	286.0	289.4	287.4	287.9	1.3153
2	288.5	285.7	289.1	287.2	287.6	1.3065
3	288.9	286.0	289.5	287.3	287.9	1.3718
4	288.8	286.0	289.4	287.3	287.9	1.3255
5	288.7	286.0	289.3	287.2	287.8	1.2903
Average of RON	288.7	285.9	289.3	287.3	287.8	—
SD	0.1356	0.1200	0.1356	0.0748	—	—
$u_1 = SD/\sqrt{n}$	0.0606	0.0537	0.0606	0.0335	—	—

$$RON_{(MC)} = 0.0708V^4 - 0.9083V^3 + 4.0292V^2 - 7.0917V + 293.3000,$$
$$(6.5)$$

$$RON_{(LSQ)} = 0.0708V^4 - 0.9083V^3 + 4.0292V^2 - 7.0917V + 292.7000,$$
$$(6.6)$$

$$RON_{(MI)} = 0.0167V^4 - 0.2333V^3 + 1.1333V^2 - 2.2167V + 288.7000;$$
$$(6.7)$$

$$RON_{(ME)} = 0.0500V^4 - 0.6500V^3 + 2.9500V^2 - 5.3500V + 289.0000$$
$$(6.8)$$

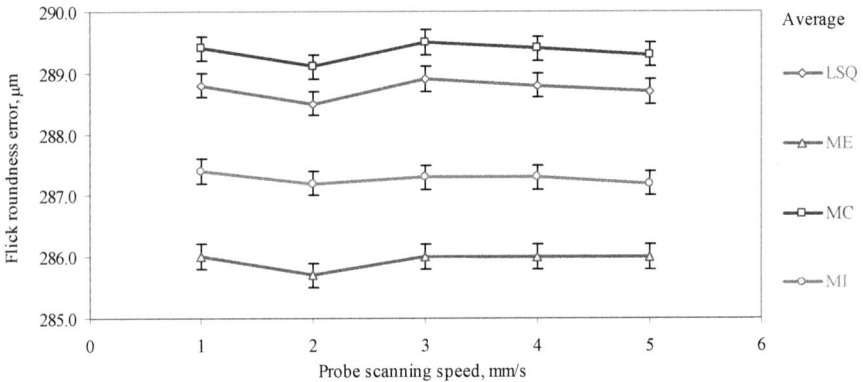

Figure 6.20 Average variation of Flick roundness measurements at different CMM strategies.

From polynomial regression Eqs. (6.5–6.8), the empirical formulae illustrate that the MC and LSQ algorithms have higher error potentials of 293.3 and 292.7 µm to the probe scanning speed, whereas the MI technique has a lowest error potential of 288.7 µm at high sensitivity coefficients of −2.2 to the probe scanning speed (V). The MC and LSQ algorithm have the same sensitivity coefficient of 4.0 to the probe scanning acceleration. The ME algorithm has an error potential of 289.0 µm at sensitivity coefficients of −5.4 to the probe scanning speed and sensitivity coefficients of 3.0 µm to the probe scanning acceleration.

6.4.3   Uncertainty Evaluation

In coordinate metrology, to measure the cylindrical artifact, as shown in our work, the measurement uncertainty mainly results from CMM machine, measurement environment, and sampling strategies [11], while the uncertainty influence of sampling strategy has not been considered here due to its complexity. The uncertainty contributions significant include the following parameters.

6.4.3.1   Repeatability

The statistical *study* based on design of *experiment* was carried out in order to *evaluate* the expanded uncertainty in measurement. Statistical study of the repeatability in measurement has been calculated and evaluated for four fitting algorithms LSQ, ME, MC and MI have the type (A) uncertainty (u_1) values of 0.0606, 0.0537, 0.0606 and 0.0335 µm, respectively, as shown in Table 6.3.

6.4.3.2   Resolution

The resolution r of CMM in last digit of a measured value, whichever is the largest, is causing an uncertainty component u_2:

$$u_2 = r/(2\sqrt{3}) = 0.1/2\sqrt{3} = 0.0289 \text{ µm}$$

6.4.3.3   Indication error

The maximum permissible error of indication is 0.9 µm. When a normal distribution is assumed, the uncertainty component is

$$u_3 = 0.9/\sqrt{3} = 0.5196 \text{ µm}$$

6.4.3.4   Temperature

The standard reference temperature for measurement is 20°C. During the implementation, the environmental temperature in the coordinate metrology laboratory at NIS was controlled within 20±0.5°C, the uncertainty component u_4 from temperature and dirt is estimated 0.05 µm.

Above components are all uncorrelated, so the uncertainties of measured points for roundness variation are calculated as follows:

$$u_c = \sqrt{u_1^2 + u_2^2 + u_3^2 + u_4^2}, \quad U_{exp} = K(u_c)$$

Eventually, the evaluation of uncertainties in this work reflects confidence in the high credibility of the proposed measurement method as shown obviously in Table 6.4. Thus, it can say that the MI fitting technique is a closer to reality and very much more accurate than the other techniques. In addition, the evaluation results illustrate that the LSQ and MC algorithms have higher uncertainties of 1.0526 µm, where MI technique has a lowest uncertainties of 1.0478 µm.

Table 6.4 Evaluation of uncertainty budget in roundness deviation measurements

CMM fitting algorithms	LSQ	ME	MC	MI
Combined standards uncertainty, u_c	0.52650	0.52555	0.52650	0.523873
Expanded uncertainty U_{exp}	1.0526	1.0511	1.0526	1.0478
Average of U_{exp}	1.0510	—	—	—

6.5 Conclusions

This chapter presents new experimental validation method to improve the measurement accuracy of dimension and roundness form of a coordinate measuring machine for particular measurement tasks. The proposed method is very important in the development issue of CMMs dynamic validation to be more accurate and precise machines in measurement. Some error formulae of two main data sets have been postulated to correlate the diameter and roundness measurements within the application range. The method proposed requires just two selections of geometric fitting algorithms and probe scanning speeds in cylindrical measurements. From this research to improve dimension and roundness measurements quality, some conclusions can be drawn:

- The new validation method of dimension and form deviation using Flick standard artifact may be a new suitable tool as a highly sensitive method for CMM performance [5].
- Suitable scanning speed of the machine touch probe should be well selected in accordance with the fitting algorithm to get high response quality with accurate dimension and roundness measurements.
- For diameter and roundness measurements, MI algorithm shows the lowest error potential indicating high quality response.
- For diameter measurement both treatment algorithms ensure stable measuring accuracy at all probe scan speeds. This is probably due to the geometric perfection resonance of the circular surface of the Flick standard artifact.
- For roundness measurement, both geometric fitting treatment algorithms ensure high measuring accuracy at 2 mm/s probe scan speed.
- For similar diameter measurements, the MI algorithm shows high quality response, while other algorithms show low quality response to the probe scanning speed.
- For similar roundness measurements, the ME algorithm shows high quality response, while the MC algorithm shows low quality response to the probe scanning speed.
- The expanded uncertainty result has been evaluated within 1.051 µm ware less than the maximum limit of the CMM permissible probing error (1.9 µm), which confirmed the high degree of confidence in the measurement results. Thus, the Flick standard artifact may be used as a simple reliable artifact for CMM evaluation.
- From the new validation method analysis, the author observed that there are proportional relationships between the uncertainty and the error potential of the empirical error formulae in measurement. Thus, the LSQ and MC algorithms have higher uncertainties, which agreed with the higher rate of there error potentials, whereas the MI technique has a lowest uncertainty of 1.0478 µm, which agreed with the lowest rate of the error potential.
- Using a Flick standard artifact should be practical high-precision incremental in dimension (diameter) and roundness

form indicator with high stability in measurement. This means a better transfer stability of CMM quality could be significantly improved.

- Eventually, the most basic measuring applications, the metrology of supplemental Flick standard artifact will increase our knowledge about the state of CMM measuring strategies.

References

1. J. Buajarern, T. Somthong, S. H. R. Ali, and A. Tonmeuanwai, Effect of Step Number On Roundness Determination Using Multi-Step Method, *International Journal of Precision Engineering and Manufacturing*, vol. 14, issue 11, pp. 2047–2050, 2013.

2. S. H. R. Ali, H. H. Dadoura, and M. K. Bedewy, Identifying Cylinder Liner Wear Using Precise Coordinate Measurements, *International Journal of Precision Engineering and Manufacturing*, vol. 10, no. 5, pp. 19–25, 2009.

3. I. Vrba, R. Palencar, M. Hadzistevic, B. Štrbac, and J. Hodolic, The Influence of the Sampling Strategy and the Evaluation Method on the Cylindricity Error on a Coordinate Measurement Machine, *Journal of Production Engineering*, vol. 16, no. 2, 2013.

4. S. H. R. Ali and J. Buajarerm, New Measurement Method and Uncertainty Estimation for Plate Dimensions and Surface Quality, Hindawi, *Advances in Material Science and Engineering*, vol. 2013, pp. 1–10, Article ID 918380, 2013.

5. H. Nouira and P. Bourdet, Evaluation of Roundness Error Using a New Method Based on a Small Displacement Screw, *Measurement Science and Technology*, vol. 25, no. 4, 2014.

6. International Standard Organization: ISO/IEC Guide 98-3:2008, Uncertainty of measurement—Part 3: Guide to the expression of uncertainty in measurement (GUM: 1995).

7. International Standard Organization: Geometrical Product Specifications (GPS)—Acceptance and reverification tests for Coordinate Measuring Machines (CMM)—Part 2: CMMs used for Measuring Size, ISO 10360-2, 2nd edition, 15-12-2001.

8. International Standard Organization: Geometrical Product Specifications (GPS)—Acceptance and reverification tests for Coordinate Measuring Machines (CMM)—Part 4: CMMs used in Scanning Measuring Mode, ISO 10360-4, 2nd edition, 15-3-2000.

9. S. H. R. Ali, Probing System Characteristics in Coordinate Metrology, *Measurement Science Review*, vol. 10, no. 4, pp. 120–129, 2010.

10. H. S. Nielsen and M. C. Malburg, Traceability and Correlation in Roundness Measurement, *Precision Engineering*, vol. 19, p. 175, 1996.

11. X. Wen, Y. Xu, H. Li, F. Wang, and D. Sheng, Monte Carlo Method for the Uncertainty Evaluation of Spatial Straightness Error Based on New Generation Geometrical Product Specification, *Springer, Chinese Journal of Mechanical Engineering*, vol. 25, no. 5, pp. 875–881, 2012.

PART 4
PERFORMANCE OF TALYROND METROLOGY TECHNIQUE

Chapter 7

Factors Affecting the Performance of Talyrond Measurement Accuracy

7.1   Introduction

Roundness is the very important feature in the quality control of mechanical products for the dynamic parts that need accurate and precise measurement. Roundness is sometimes called circularity. The purpose of studying roundness form deviation of circle and cylindrical features is to avoid the excessive lateral or axial runout deviation of the rotating and reciprocating parts during dynamic operation. After production procedures, manufacturing metrology of roundness feature can be done using advanced measuring instruments. The roundness geometrical machine describes the condition on a surface of revolution where all points of the surface intersect.

Historically, roundness measurement was based on the use of some simple tools such as dial indicator. However, after the industrial revolution appeared the roundness measuring instruments that based on using one of two types of machines. The first one is called Talyrond machine. Another way to measure surface roundness is to use a coordinate measuring machine (CMM). The roundness measurement technique using Talyrond instrument is based on one of the two versions of configuration. The configurations of Talyrond technique have either rotating

Automotive Engine Metrology
Salah H. R. Ali
Copyright © 2017 Pan Stanford Publishing Pte. Ltd.
ISBN 978-981-4669-52-8 (Hardcover), 978-1-315-36484-1 (eBook)
www.panstanford.com

table or rotating spindle (hydrostatic). The most common type is the hydrostatic spindle configuration version. The rotating pick-up version of the instrument was first made; this was termed Talyrond-1, which was developed later. The ultra high precision instrument as Talyrond HPR (TR-73) machine becomes one of the important tools in precision engineering metrology. The standard TR-73 has three accurate orthogonal axes and is equipped with high sensitive touch probe. The TR-73 probe cantilever is brought into contact with the inner or outer circular surface of object being measured at a recorded position. In the measurement operation, the probe stylus profile senses the surface height through mechanical contact, while the stylus traverses the peaks and valleys of the surface with very small contacting force. The horizontal motion of the stylus tip is converted to an electrical signal by a transducer. A number of points are taken around the component and these are then combined in computer software to determine the accurate roundness profile of the object, which represents the actual surface profile. Therefore, the Talyrond stylus system is directly sensitive to surface height with little interference.

The deviation in roundness metrology using TR-73 instrument can be related to the performance and experience of operator, environmental interaction, workpiece finishing, and accuracy of software certainly. It can be assumed that some influence parameters of operator behavior and TR-73 software strategy techniques have effective reactions on the measurement quality assurance [1–2]. The measurement accuracy of a work piece influences by many different metrological parameters. The error characteristics in the TR-73 software are very important from the metrological point of view to find an optimum fitting solution. The TR-73 software data analysis can contribute significantly to the roundness measurement accuracy of measured object.

This chapter presents the work to develop a testing algorithm and evaluation program as a metrological strategy for roundness measurement. In this work program, the fitting software techniques equipped with 10 different strategy parameters using TR-73 machine are studied experimentally. These parameters included two types of fitting filters and four types of software algorithm techniques during four ranges of spectral wave numbers using undulations per revolution (upr). This work program will be available as an activity provided by the National Institute for Standards (NIS), Egypt. The main objective is to eliminate the reputable errors in turning operator during measurement because they direct impact form metrology. The goal is to reduce costs according to consuming measurement time and improve figure accuracy of visible roundness measurement. Consequently, the advantage of the research is that it helps the TR-73 operator in developing a methodology for precision assembly as well as an error compensation method to improve the overall accuracy. This study is very important for the TR-73 software designer to develop new version of precision machines. It confirmed that the Talyrond-73 device as an ultra high precision machine is a powerful tool for roundness form metrology in modern engineering industries [3].

7.2   Background and Motivation

Great benefits of manmade objects have been achieved, from the alphabet to wheel invention passing rotating parts. The final quality of an engineering rotating product as used in an airplane and a motor vehicle is influenced by various factors [2–4]. The required tolerances for dynamic rotating parts continuously have to be very fine, whereas the complexity of work pieces increases. However, the selected technique for measurement has to be applied in order to achieve precision measurements and sufficient accurate results. The measurement quality of the TR-73 instrument can be dependent on the operator behavior, environmental interaction, work piece finishing, and machine accuracy. It can be assumed that some influence factors of operator behavior and TR-73 machine software accuracy have an effect on the measurement quality. There are two disadvantages of the stylus instruments in that the probe tip may damage the scanned surface (depending on the hardness of the surface relative to the stylus normal force) and the stylus tip size [3]. However, the Talyrond TR-73 machine is different in that the force exerted by the touch probe tip on the sample surface is very small, up to less than 1 N. Moreover, the author and other researchers find that it is difficult to separate the error resulting from the stylus vibration and the result of measured surface in roundness measurement [4–6].

The roundness feature means the change in the radius of an object and is usually referred to as "circularity" or "out-of-roundness." The measure of roundness is expressed as the difference between the smallest and largest diameters, which can be expressed by peak and valley in metrology. To assist this measurement, a mathematical reference circle is used through machine software. The position of the reference circle to the measured profile and its center is not arbitrary and should be selected by the metrology engineer to ensure the measurement result that meets the required specifications [7]. The software package of the Talyrond machine includes specific important strategic techniques of measurement to be used in roundness assessment.

7.2.1 Fitting Filters

Another important parameter in the TR-37 machine during measurement is the fitting filters. Generally, fitting filters are used to separate the number of lobes exiting a component's measured surface. These are used to eliminate the sources of error and to avoid problems such as swarf machining. They are also used to remove the unwanted noise signal from the measurement results. Noise often results from vibration dynamics in the machining process and from other processes or from the measurement instrument itself. Moreover, filters are used for separate low-frequency data from high-frequency data signals to simplify the evaluation of measurement results. Filters are also used for separate lobbing frequencies to evaluate the process and to assure highly reliable measuring results.

7.2.2 Fitting Spectral Wave Responses

One important parameter in the TR-37 machine based on the operator use is the suitable selection of spectral wave numbers in measurement strategy. The frequency undulations per revolutions are the number of surface-profile deviations from a true circle in one revolution. Electronic filters with different frequency responses of spectral waves are used to evaluate the surface profile. Standard filter responses are used in the ranges of 1 to 15, 1 to 50, 1 to 150, and 1 to 500 upr. Figure 7.1 illustrates the electronic ranges of fitting filters with varying frequency responses are used to evaluate the roundness feature in measurement. For example, a 1 to 50 filter removes undulations above 50 per revolution from the measured feature profile. Without filtering, the high frequency waves can make evaluating the out-of-roundness difficult because they can conceal the lower frequencies. Often the low-frequency waves are of larger amplitude and thus of greater importance to the manufacturer.

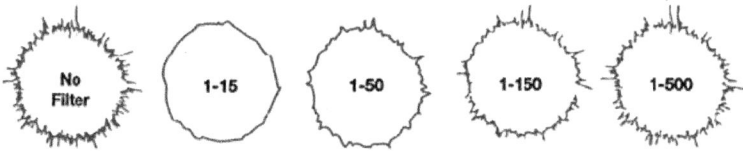

Figure 7.1 Electronic ranges of fitting spectra wave numbers responses in the roundness profile.

7.2.3   Fitting Algorithms

Fitting algorithm is one of the most important themes in modern dimensional metrology machines. Fitting algorithm software is particularly true in coordinate measurement systems such as CMMs and Talyrond machine. It works based on computations and convert the collected raw data to report results can be a major source of error in measurement techniques. The role of geometric fitting for roundness feature is to reduce measured point coordinates to curve and surface parameters. The resulting curves of surface are called the substitute geometry for the roundness measurement. In further processing, the computed parameters are compared to the tolerance limits for the measured part. Thus, it was found that the error estimation of the computed substitute geometry is important to determine the quality of a measurement. There is no obligatory standard or accepted method for evaluating the effects of fitting software on the error estimation for Talyrond measurements. ISO/TS 12181-1 and 12181-2 standard defines four mean specific reference fitting software methods [8–9]. Figure 7.2 shows the basic idea of four internationally defined fitting algorithm techniques of circles that are used in roundness measurement in National Metrology Institutes (NMIs) and industry today [2, 10]. Least square circle (LS) is well suited for establishing a datum axis; maximum inscribed circle (MI) is the largest true circle that will fit inside the measured profile; the minimum circumscribed circle (MC) is the smallest true circle that will contain the measured profile; and minimum zone circles (MZ) are two concentric circles having the same center, which enclose the measured profile and have a minimum radial separation.

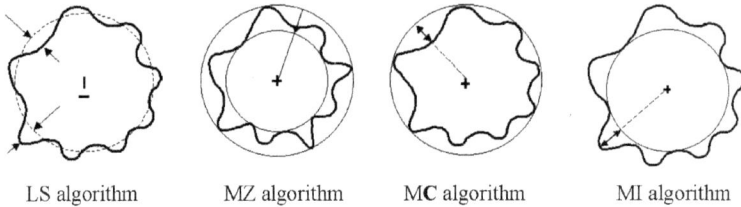

LS algorithm   MZ algorithm   MC algorithm   MI algorithm

Figure 7.2   Reference software fitting circles used in roundness measurement.

7.2.4   Types of Errors

In fact, machined engineering surface does not have perfect form due to various error sources such as machining process and finishing quality. Additional parameters likes measuring instrument accuracy and strategy of measurement besides human and environmental conditions are common types that affect the result of measurement errors. The purpose of TR-73 fitting software strategy is to determine the final fine feature of the object. There are many different sources of errors which may influence roundness measurements [11–15]. Some typical sources of these errors are hysteresis sensitivity, probe tip size, contacting force, spindle motions (axial, radial and tilt), alignment of artifact, thermal drift, closing error, electrical noise, mechanical vibration and work-piece contamination [16, 17]. The standard uncertainty budget of a roundness measurement using TR-73 instrument has been estimated within 1.3 nm [12]. Another research work estimates the combined uncertainty in measurement at 95% certainty to be 8.01 nm [13]. Consequently, in this work, the sources of roundness errors will be studied and analyzed to avoid two main practical errors in measurement: *instrumentation error* and *human error*. These two common types of errors with some others exist in all real measurement scenarios. The *error* generates at use *instrumentation* when data points are collected due to scanning of an object surface. TR-73 hardware adaptation induces different sources of error based on spindle error, probe system and transmutation system imperfections, leading to some inaccuracies in the measured points. Measurement error in the measuring process comes mainly from curve fit and computational resolution based on software adaptation. The *human error* arises when the operator prefers impossible strategic parameters in measurement of an object. Therefore, TR-73 operator skills and behaviors have a direct significant effect on the roundness error in future measurement. This research will provide an analytical study to avoid the sources of errors which have direct impact on form metrology in order to improve the quality of roundness accuracy in measurement method.

7.3   Experimental Work

Evaluation of the Talyrond HPR fitting software strategies through sample carrying out signals has been performed experimentally. The measurement evaluation process includes four basic steps of the instrumentation system: experimental verification of stylus contact, a data generator, reference algorithm, and a comparator to analyze and interpret the monitoring results. The TR-73 strategies in measurement have two types of fitting filters and four fitting algorithm categories at four different spectral wave responses. The software filters types were Gaussian and 2CR filters for data facilitation of circular, spherical, and cylindrical measurements. The software fitting algorithm techniques of roundness form measurement were least square (LS), minimum zone (MZ), minimum circumscribed (MC) and maximum inscribed (MI) techniques. While the response of spectral wave numbers of machine software includes dominant harmonics, wave ranges from 1–15, 1–50, 1–150 to 1–500 upr have been used. The machine software strategy and stylus scanning speed were selected and primary tested in the recommended environmental conditions. Centering object seat on machine table base was finally cleaned and located on the test position. The measured object is standard accurate cylindrical circular surface. The TR-73 machine was turned on to check the electric power switches, hydrostatic-bearing spindle rotation, and stylus speed, where a Hatchet styles tip of the long type has been selected and calibrated according to the machine working manual. The performance of the TR-73 accuracy in scanning measuring mode has been verified and accepted within standard specification according to ISO/TS 12181-1/2 [8–9].

Inspection of cylinder feature consists of measuring and presenting surface elements. To determine the approximation accuracy of the fitting algorithms, a geometrical form has been created. Relevant influences in the roundness accuracy measurement have been taken into account according to standards. The development of software analysis tools for fitting strategy parameters and their validation is another major challenge in this work. The specifications of stylus and TR-73 test machine are presented in Table 7.1. Figure 7.3 illustrates the typical monitoring storage results of roundness measurement using datum spindle. The preliminary result shows that the roundness measurement in Fig. 7.3b is substantially enhanced compared with that shown in Fig. 7.3a, due to fitting strategy parameters.

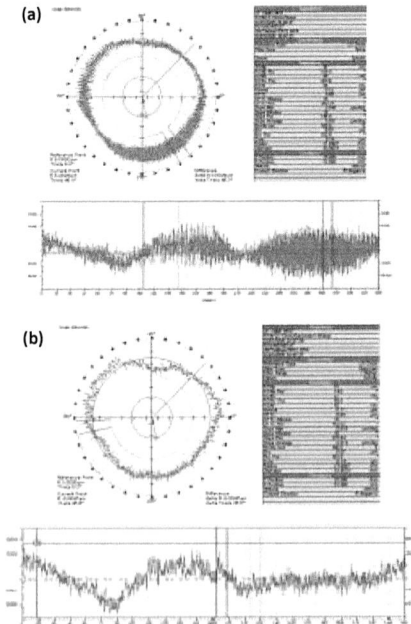

Figure 7.3 Typical result of roundness measurement before and after fitting parameter using TR-73 instrument. (a) Roundness measurement with non-filter. (b) Roundness measurement using Gaussian filter.

Table 7.1   Data specification of TR-73 machine and stylus

Software code no.	: M 112/2266-02
Measurement direction	: Anti-clockwise
Attitude	: Vertical
Stylus no.	: K42/3827 TR73 1.27 mm Hatchet
Measurement speed	: 6.0 rpm
Angle range trace	: 360°

Experimentally, the procedures of roundness feature measurement, the software filters, circle fitting techniques, and spectral wave numbers have been studied for each measurement strategy. The determination of roundness errors has been studied to differentiate the deviations between the evaluated parameters of all fitting software strategies. The analysis of roundness measurement accuracy in this research has been done to predict the effect of spectral wave numbers at different specific software parameters of measurement strategy as follows:

- roundness measurement at two different fitting filters
- roundness measurement at four different fitting techniques with Gaussian filter
- roundness measurement at four different fitting technique with 2CR filter.

7.4   Results and Discussion

7.4.1   The Effect of Fitting Filters

Peak and valley signals (RON_P; RON_V) in roundness metrology are the basic parameters to study the total error of signal profile. The total measurement error of roundness (RON_t) is the distance of a form profile between highest peak-to-valley response signals. The effects of computational geometric software fitting filters on the roundness feature errors have been measured. Thus, depending on the separation process of signals that has been installed using the LS fitting algorithm technique in the rest of the roundness measurements. The effects of two types of software filters on the peak and valley signals have been presented in

Table 7.2 to compare the deviations in measurement accuracy. Figure 7.4a shows the effect of Gaussian-, 2CR- and none-filter on the peak and valley response using LS fitting algorithm at different spectral wave numbers. It is illustrated that whenever spectral wave numbers increase in the measurement, the error in the roundness measurement increases for each filter. However, the RON_t response does not exceed 49 nm despite a change of spectral wave numbers when the filters are not used. Therefore, the results using LS fitting algorithm confirm that the use of the Gaussian filter gives minimum error response in roundness measurement within the application range.

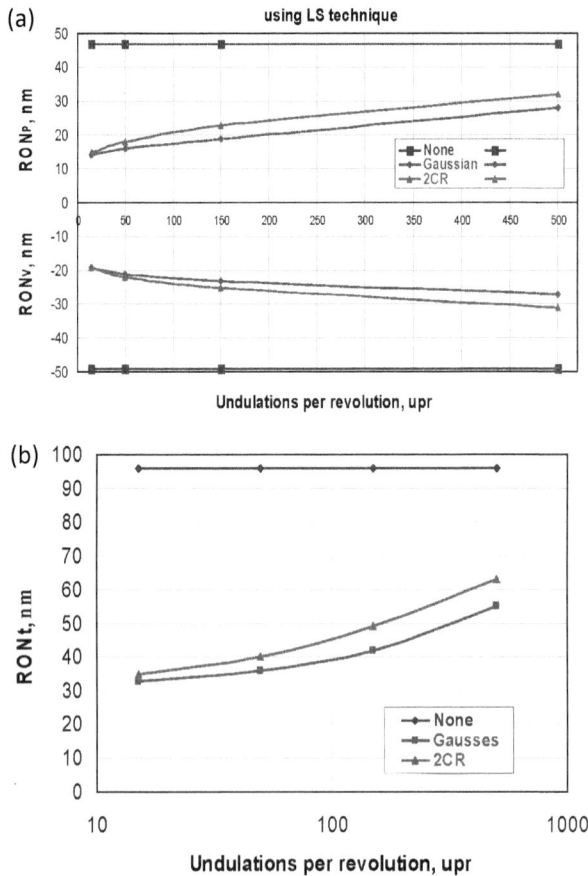

Figure 7.4 The influence of spectral wave numbers using Gauss-, 2CR-, and none-filter at (a) RON_P; RON_V and (b) RON_t.

Table 7.2 Roundness measurement errors of fitting filters at LS algorithm technique

Frequency	Peak roundness error, RON_P (nm)			Valley roundness error, RON_V (nm)		
	None	Gaussian	2 CR	None	Gaussian	2 CR
1–500		28	32		27	31
1–150	47	19	23	49	23	25
1–50		16	18		21	22
1–15		14	15		19	19

Figure 7.4b shows the effect of computed software filtering on the RON_t output signals using Gaussian-, 2CR-, and none-filters at spectral wave numbers changes with fixing LS fitting algorithm. It is noticed that with increased upr, the RON_t increases using each filter. However, the roundness total error is almost constant despite a change in upr if no filter is used (blue). The analysis of the results confirmed that the use of the Gaussian filter gives the lowest error of RON_t measurement at the same conditions, while the impact of fitting filters and fitting techniques on roundness accuracy development still needs more analysis to establish a reference data set in these measurements.

7.4.2 The Effect of Gaussian Filter and Fitting Techniques

Because the error response signal using Gaussian filter has the lowest computational geometric error in roundness measurement when using 2CR filter (Fig. 7.4), this guides the author to install a Gaussian filter in the following measurements in this section. Therefore, the computational effect of Gaussian filter on the roundness feature signal at different types of fitting techniques has been measured. The cleared outputs of Gaussian filter response in the peak and valley (RON_P; RON_V) signals have been measured and registered in Table 7.3. Figure 7.5 shows the impact of Gaussian filter on the peak and valley signals of roundness error responses using four fitting algorithm techniques with variable upr. It is illustrated that the increased of spectral wave numbers leads to increased error in measurement for both peak and valley in each fitting, while the peak and valley error of roundness profile is almost constant despite a change in upr if no filter is used. This confirms that the use of the MC fitting technique at Gaussian filter response gives lower error in the peak roundness measurement in certain conditions, while using the MI technique at Gaussian filter response gives lower error in the valley roundness measurement in certain conditions.

Table 7.3   The roundness errors (nm) using Gaussian filter type at various algorithms

	Peak roundness error, RON_P (nm)				Valley roundness error, RON_V (nm)			
Frequency	LS	MZ	MC	MI	LS	MZ	MC	MI
500	28	24	1	51	27	24	66	1
150	19	18	1	38	23	19	51	0
50	16	15	1	36	21	17	45	1
15	14	13	1	27	19	14	40	0

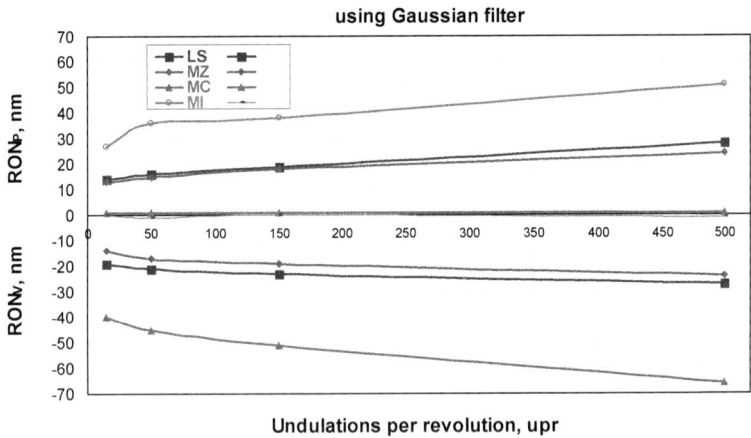

Figure 7.5   The influence of spectral wave numbers on the roundness error at various algorithm techniques.

The computation effects of Gaussian filter on the total errors (RON_t) have been registered in Table 7.4 and shown in Fig. 7.6. It is shown that the minimum geometric error was while using the MZ technique, while the maximum response was while using the MC technique of measurement. Therefore, the results illustrated that if the metrologist selects the MZ fitting algorithm, as shown in Fig. 7.6, the use of the Gaussian filter response gives the lowest error in the roundness measurement in certain conditions. These results will be a guide for the metrologist using the Talyrond HPR machine.

Table 7.4 Total error form of roundness with various algorithms at Gaussian fitting type

Frequency	Total error of roundness, RON_t (nm)			
	LS	MZ	MC	MI
500	55	48	67	52
150	42	37	52	38
50	36	32	46	37
15	33	27	41	27

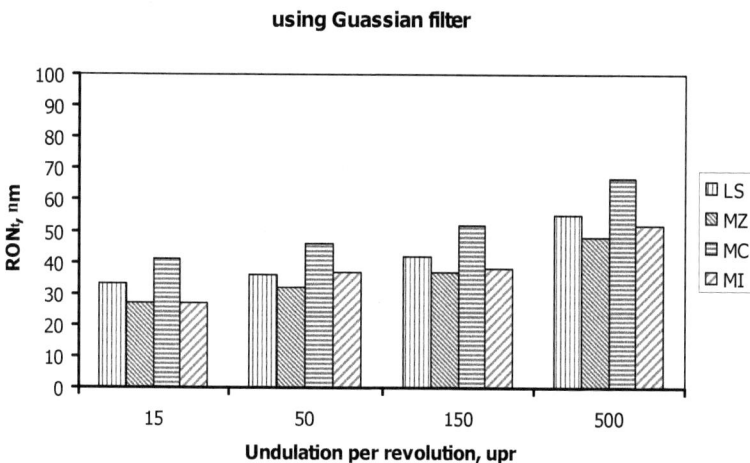

Figure 7.6 Total error of roundness for various algorithm techniques at different spectral wave numbers.

7.4.3 The Effect of 2CR Filter and Fitting Techniques

The peak and valley signals and total computed error in roundness feature profile using 2CR filter response have been measured and registered in Tables 7.5 and 7.6. It is clear that using the MC technique with 2CR filter response gives lower error in the peak signal, while using the MI technique gives lower error in the valley roundness measurement. However, the lower error in the total roundness measurement was achieved while using the MZ fitting technique in certain conditions.

Table 7.5 Roundness errors at various fitting algorithms using 2CR fitting type

Frequency	Peak roundness error, RON_P (nm)				Valley roundness error, RON_V (nm)			
	LS	MZ	MC	MI	LS	MZ	MC	MI
500	32	30	2	68	31	30	68	0
150	23	21	0	44	25	21	59	0
50	18	17	1	35	22	17	48	2
15	15	15	1	29	19	15	42	0

Table 7.6 Total roundness value (nm) with four fitting algorithms at 2CR filter type

Frequency	Total error of roundness, RON_t (nm)			
	LS	MZ	MC	MI
500	63	61	70	68
150	49	42	59	44
50	40	35	49	37
15	35	30	43	29

Figure 7.7 shows the influence of 2CR filter response on the peak and valley signals of roundness error using four fitting algorithms at spectral wave numbers changes. It is illustrated that any increase in frequency leads to an increase in the roundness error in both peak and valley in each fitting. The minimum computational error of geometric measurement was observed while using the MZ technique, while the maximum response technique was observed while using MC in both peak and valley positions in roundness measurement. Thus, it can be said that if the metrologist selects the MZ fitting algorithm technique, as shown in Fig. 7.8, it confirms that the use of the 2CR filter gives lower computational error in the roundness measurement in the same conditions. This result helps the metrologist use a suitable reference data set for circular measurement using the 2CR fitting technique on the Talyrond HPR machine.

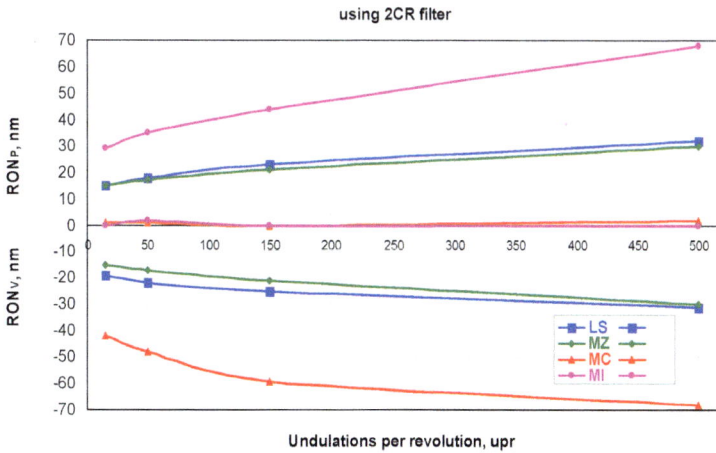

Figure 7.7   Illustrate the effect spectral wave numbers on the roundness variations.

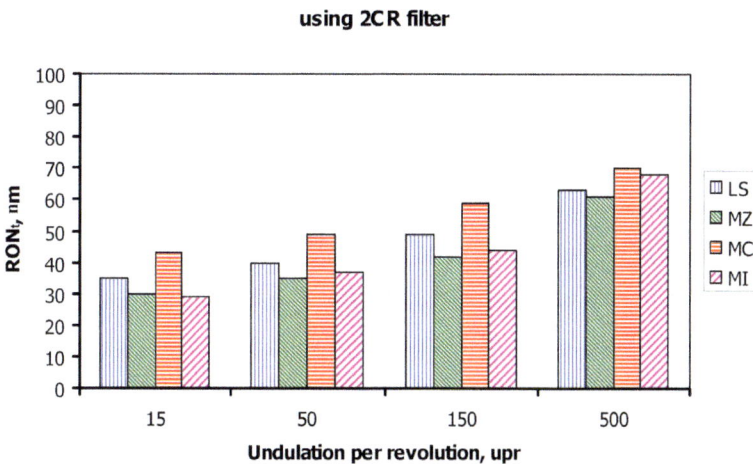

Figure 7.8   The RON_t using four algorithm techniques at different spectral wave numbers.

7.5   Analysis and Estimation of Roundness Accuracy

The computational geometric effects of peak, valley, and total errors in roundness metrology using LS fitting technique have been analyzed. The relative results of these errors are presented in Tables 7.7–7.9. Table 7.7 shows the rate of improvement of the peak roundness error signals (RON_P) while using Gaussian- and 2CR- filters and none-filter when fitting algorithm is the LS technique. Table 7.8 shows the rate of improvement in valley roundness error signals (RON_v). Table 7.9 presents the achievement rate of RON_t in certain conditions. Thus, it can be stated that if the metrologist selects a suitable measurement strategy with Gaussian filter while using the LS fitting technique, it gives lower computational errors for RON_P, RON_v, and RON_t in measurement, especially at 15 upr using the Talyrond HPR machine.

Table 7.7 Improvement rate of peak roundness accuracy using two filters at LS fitting algorithm

Frequency	**Improvement in RON_P measurement**		
	None	**Using Gaussian, %**	**Using 2 CR, %**
500		59.6	68.1
150	47.0	40.4	48.9
50		34.0	38.3
15		29.8	31.9

Table 7.8 Improvement rate of valley roundness accuracy using two filters at LS fitting algorithm

Frequency	**Improvement in RON_v measurement**		
	None	**Using Gaussian, %**	**Using 2 CR, %**
500		55.1	63.3
150	49.0	46.9	51.0
50		42.9	44.9
15		38.8	38.8

Table 7.9 Improvement rate of total roundness accuracy using two filters at LS fitting algorithm

Frequency	**Improvement in RON_t measurement**		
	None	**Using Gaussian, %**	**Using 2 CR, %**
500		57.3	65.6
150	96.0	43.8	51.0
50		37.5	41.7
15		34.4	36.5

The experiments revealed that the maximum roundness accuracy improvement could be high, with 68% achieved at 2CR fitting filter, while the average roundness accuracy is found to improve by about 59.6% using the Gaussian fitting technique. Moreover, the average value of roundness accuracy could be high, achieved as 63.3% at 2CR filter, while the average roundness accuracy is found to be improved by about 55.1% using the Gaussian fitting technique, which proved the effectiveness within the application range.

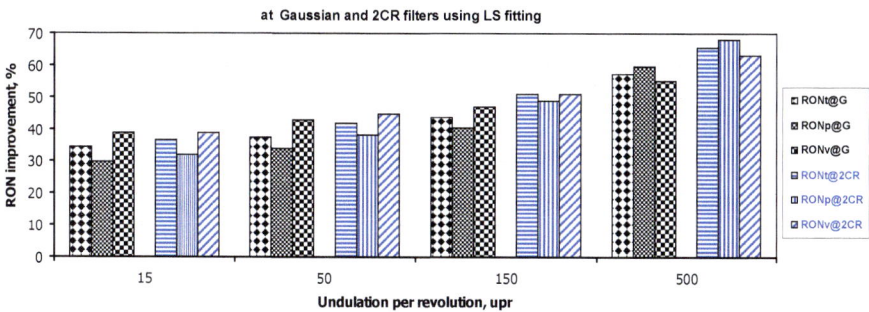

Figure 7.9   Illustrate the effect spectral wave numbers on the RON_t, RON_P and RON_V accuracy variations.

7.6   Conclusion

This work carries out the data analysis of software algorithm techniques to achieve improvement in roundness measurement accuracy. New experimental investigation is made to establish a computational geometric set to improve the roundness accuracy of the Talyrond 73 HPR instrument. The experimental methodology proposed involved a detailed study of 10 main parameters at certain conditions, as they have a direct impact on form metrology. The results showed that the accuracy of roundness form measurement of object feature is a great challenge for the metrologist, especially in NMIs. The following conclusion can be drawn:

(1) There are error differences of roundness metrology in the same artifact detecting circle; this is due to the selection of different fitting algorithms, fitting filters, and frequency

range in measurement, which have different responses according to software design within the maximum permissible error.

(2) The deviation in the measurement error of roundness is not affected by the number of spectral waves in the absence of any filter.

(3) Deviation in roundness result increases with an increase in the spectral wave numbers with the use of any type of filters and may be because of the response impact of the filter design.

(4) The accuracy improvement in roundness measurement is achieved when using the Gaussian filter instead of the 2CR filter. Of course, both filters give a better response than in absence of any filter.

(5) Improvement in roundness measurement accuracy when using the MZ fitting technique is achieved compared with the MC technique, more than the MI technique, respectively.

(6) There is great variation in the accuracy of measurement for roundness feature up to twice using fitting strategy parameters from 1–15 to 1–500 upr.

(7) The results clearly indicate that the contact stylus Talyrond HPR (TR-73), as an ultra-high-precision machine and one of the advanced nanometrology machines, is a powerful tool to get the object feature requirement of roundness measurement for NMIs traceability, professional tests, and modern engineering industries.

References

1. C. Diaz and T. H. Hopp, Testing of Coordinate Measuring System Software, Proceedings of the American Society for Quality Control (ASQC), Measurement Quality Conference, National Institute of Standards and Technology (NIST), Gaithersburg, MD, 26–27 October, 1993.

2. S. H. R. Ali, The Influence of Fitting Algorithm and Scanning Speed on Roundness Error for 50 mm Standard Ring Measurement using CMM, *Metrology and Measurement Systems*, vol. XV, no. 1, pp. 33–53, 2008.

3. S. H. R. Ali, Advanced Nanomeasuring Techniques for Surface Characterization, *International Scholarly Research Network, Optics Journal*, vol. 2012, ID no. 859353, pp. 1–23, 2012.

4. S. H. R. Ali, Novel Prediction Analysis Method for Error Separation of Stylus System and CMM Machine, *Advanced Materials Research*, vol. 875–877, pp. 671–679, Feb 2014.

5. A. Nafi and J. R. R. Mayer, Probing Variance Separation on Coordinate Measuring Machines, *Proceedings of the World Congress on Engineering (WCE)*, London, UK, pp. 6–8 July 2011.

6. S. H. R. Ali, Probing System Characteristics in Coordinate Metrology, *Journal of Measurement Science Review, Institute of Measurement Science, Slovak Academy of Sciences*, vol. 10, no. 4, pp. 120–129, 2010.

7. C. J. Li and S. Y. Li, On-Line Roundness Error Compensation Via P-Integrator Learning Control, Columbia University, New York, 10027, USA, 1991.

8. ISO/TS 12181-1: Geometrical Product Specifications (GPS)—Roundness; Part 1: Terms, Definitions and Parameters ofRoundness, 2003.

9. ISO/TS 12181-2: Geometrical Product Specifications (GPS)—Roundness; Part 2: Terms, Definitions and Parameters of Roundness, 2003.

10. W. Sui and D. Zhang, Four Methods for Roundness Evaluation, *Physics Procedia*, vol. 24, pp. 2159–2164, 2012.

11. D. Chen, J. Fan, and F. Zhang, An Identification Method for Spindle Rotation Error of a Diamond Turning Machine Based on the Wavelet Transform, *International Journal Advanced Manufacturing Technology*, vol. 63, pp. 457–464, 2012.

12. R. Thalmann, Basics of Highest Accuracy Roundness Measurement, Simposio de Metrología, 25–27 October 2006.

13. S. Piengbangyang, T. Somthong, J. Buajarern, and A. Tonmueanwai, Roundness Measurement Capability and Traceability at NIMT, XIX IMEKO World Congress, Fundamental and Applied Metrology, Lisbon, Portugal, September, 6–11 2009.

14. A. Janusiewicz, S. Adamczak, W. Makieła, and K. Stępień, Determining the Theoretical Method Error during an On-Machine Roundness Measurement, *Measurement*, vol. 44, no. 9, pp. 1761–1767, 2011.

15. K. Stępień and W. Makieła, An Analysis of Deviations of Cylindrical Surfaces with the Use of Wavelet Transform, *Metrology and Measurement Systems*, vol. XX, no. 1, pp. 139–150, 2013.

16. O. Jusko, H. Bosse, D. Flack, B. Hemming, M. Pisani, and R. Thalmann, A Comparison of Sensitivity Standards in Form Metrology: Final Results of the EURAMET Project 649, *Measurement Science and Technology*, vol. 23, pp. 1–7, ID no. 054006, 2012.

17. S. H. R. Ali and H. Sediki, CNC-CMM Measurement Accuracy and Accompanied Uncertainty at Different Alignment Positions of Long GBs, *International Review of Automatic Control (I.RE.A.CO.)*, vol. 7, no. 5, pp. 485–491, 3 Nov. 2014.

PART 5
METROLOGY IN AUTOMOTIVE ENGINES

Chapter 8

Metrology as an Inspection Tool in New or Overhauled Water-Cooled Diesel Engines

8.1 Introduction

The inspection of mechanical parts designates the processes of measuring their dimensional and geometrical features. The purpose of these measurements is to assess and ensure the compliance of the parts with the intended design specifications. Particularly in the automotive applications, dimensional and geometrical inspection constitutes a dominant portion of total inspection work [1–12]. Figures 8.1 and 8.2 illustrate the basic terminology of the engine constituting parts and the four stroke cycle operation [7, 8]. The neighboring engine parts with mating surfaces in relative motion, either reciprocating or rotating, are considered the most vital parts as they are practically responsible for the friction losses, energy dissipation due to friction resistance, as well as the associated wear occurrence. The wear is a phenomenon which is accelerating by nature due to simultaneously interacting different complex tribological mechanisms during operation. It results in dimensional and geometrical distortion on those working surfaces and subsequent continuous energy and pressure languish in the engine to the limit of urgent demand for an overhaul.

Automotive Engine Metrology
Salah H. R. Ali
Copyright © 2017 Pan Stanford Publishing Pte. Ltd.
ISBN 978-981-4669-52-8 (Hardcover), 978-1-315-36484-1 (eBook)
www.panstanford.com

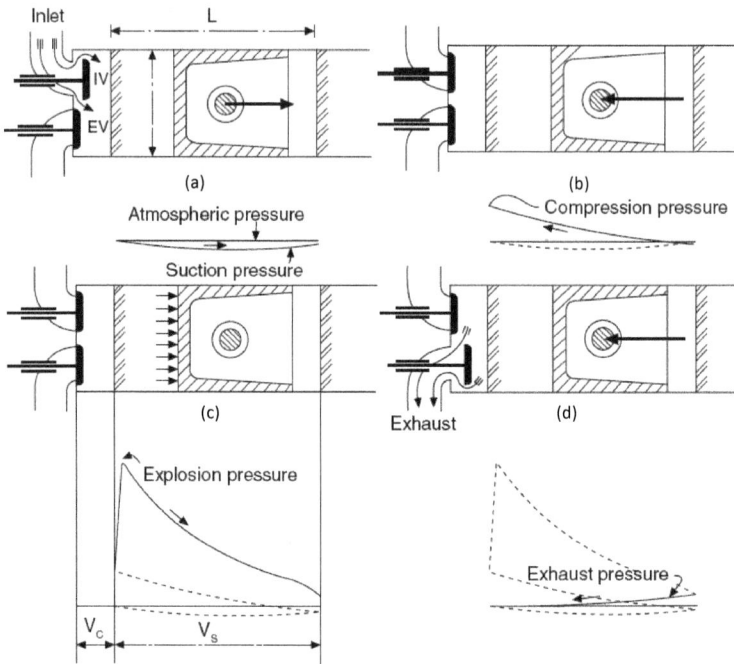

Figure 8.1 Four stroke cycle engine [7].

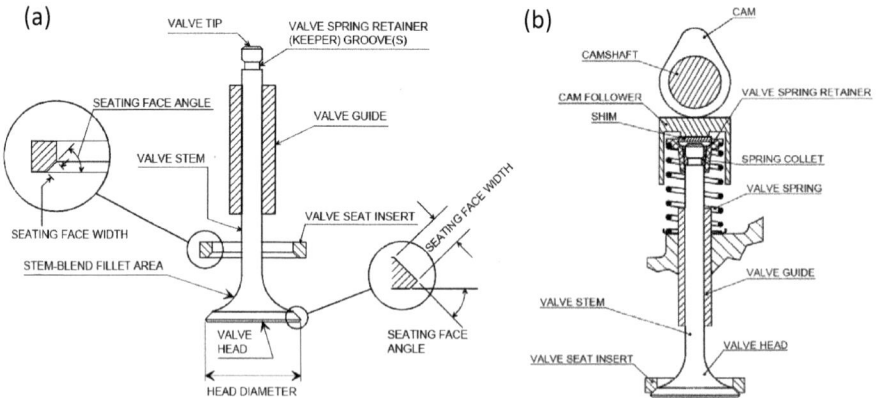

Figure 8.2 Engine valve mechanism [8]. (a) Valve and seat insert. (b) Overhead camshaft valve drive.

This chapter discusses the decisions originally taken by the engine designer regarding the dimensional, geometrical, and tolerances of the contacting working surfaces of the neighboring parts, type of fit, materials of those parts, and their surfaces treatments would markedly affect the reliability, durability, performance, and life of the engine [4, 6–9]. Thus, the various metrology equipment and tools available for the dimensional and geometrical features measurements play an important role in the inspection and testing processes of automotives. However, there are some other factors relating to operational conditions and quality of manufacturing that would also affect the engine performance characteristics and service life cycle duration [8]. Therefore, many research works have been conducted on dimension precision measurements and surface inspection techniques with the objective of reverse engineering applications and/or automotive engine performance assessment [1–9].

This chapter describes a developed program to inspect and investigate crucial engine parts during an overhaul using advanced metrology measurement devices such as CMM, and CCD camera. CNC-CMM is an advanced multi-purpose coordinate measurement system that helps keep pace with the modern production requirements. It replaces prolonged, complex, and inefficient conventional inspection methods. CMM provides instant measurement results using special setup and operating procedures by a special metrologist. It can be used to check dimensional and geometrical accuracy of almost every configuration from small rings, to engine blocks, and even to circuit boards [6]. Common Zeiss optical microscope equipped with charge-coupled device (CCD) camera for pixel size measurement technique [13, 14] is used to measure the critical lapped contact area between valve and valve seat. The engine performance characteristics have been evaluated and correlated to the parts surfaces measured parameters resulted from the inspection program. Uncertainty of the measurements has been assessed and analyzed.

8.2 Engine Inspection Program

Precise determination of the dimensional and geometrical deviations from the specifications put down by the engine designer dictates careful development of inspection program for the quality and suitability assurance of engine crucial replaceable parts. Successful inspection program should conclude the following: first, identifying the crucial replaceable parts from which the *cylinder liner, piston, valves*, and *valve seats* were adopted throughout this work. Second, proper selection of precise available measuring devices from which CMM, CCD camera equipped optical microscope, and cylinder compression pressure indicating device were verified and employed. In addition, exact locations and/or paths of measurements on each element together with detailed measurements implementation sequence and procedures should also be clearly specified according to configuration and size.

8.2.1 Engine General Specifications

The overall specifications of the overhauled engine under investigation used to demonstrate the developed inspection program in this work are tabulated in Table 8.1.

Table 8.1   Specifications of the engine in overhaul

Engine model	Diesel, 4 cylinders, in line
Engine capacity	2200 cm^3
Mileage at overhaul	249949 km (1998–2008)
Nominal size of the new piston	85.4 mm
Nominal size of the new liner	86.0 mm

8.2.2 Cylinder Liner Inspection

Locations and paths of CMM measurements conducted on the bore of replaced cylinder liners newly fitted in the cylinder block are depicted in Fig. 8.3. Both roundness and straightness geometrical parameters were measured along the indicated longitudinal and circular paths. Figure 8.4 illustrates the polar coordinates of the center location (r and θ) as transformed from the Cartesian coordinates (x and y) raw data measured at the 12 transverse circular paths indicated in Fig. 8.3a—in other words, along the axis of each cylinder liner. Thus, coaxiality in each liner could be evaluated and compared with the others.

It is worth mentioning that the interval deviations along the Cartesian coordinate (z) measurements have been found negligible.

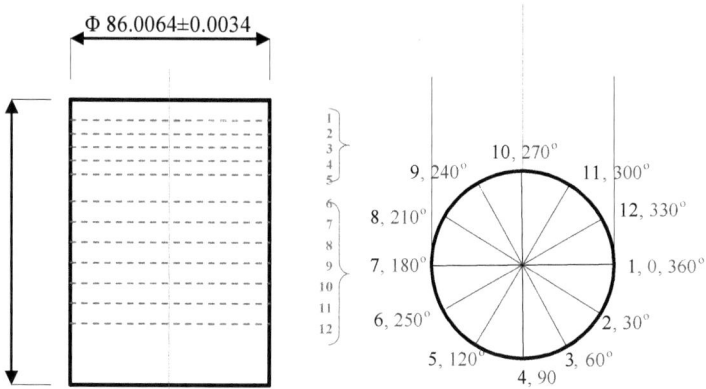

a) Locations for roundness measurements. b) Locations for straightness measurements

Figure 8.3 Specified measurements locations on cylinder liners.

The CMM measurements of total roundness have been exploited to simulate the bore surface topography and roughness represented by the mean peak and mean valley in addition to the mean peak to valley values as illustrated in Fig. 8.4 and Eq. (8.1).

$$RON_t = RON_P + RON_V \qquad (8.1)$$

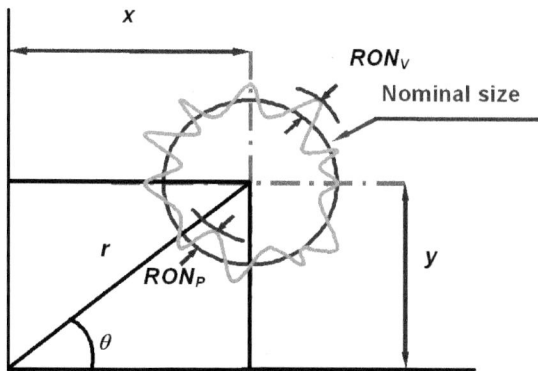

Figure 8.4 Transformation of Cartesian coordinates of the centers into polar ones.

8.2.3 Valve Lapped Area Inspection

Contact surface areas between inlet and exhaust valves and their seats shown schematically in Figs. 8.5 and 8.8 have been evaluated from measurements carried out by an optical microscope equipped with CCD camera. This method is based on the pixel size measurement technique as an intrinsic optical signal [13, 14]. The digital image processing was compared to optical reference standard scale.

The diameter (d_2 and d_3) values were measured by the CMM while the magnitude of (a) was experimentally evaluated by the CCD camera using the optical pixel image size technique for both inlet and exhaust valves and seats. The contact area (A_c) of the lapped portion of surfaces between the valves and the seats could be computed by applying the following simply derived trigonometric relation Eq. (8.2):

$$A_c = \frac{\pi a}{2}(d_2 + d_3) \tag{8.2}$$

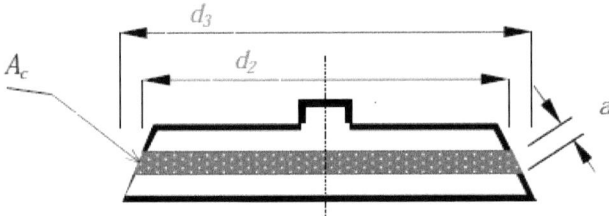

Figure 8.5 Schematic of the lapped area.

8.2.4 CMM Verification

The maximum permissible error according to the ISO 10360 standard [16] of the used CMM machine can be judged using the following equations:

$$\mathrm{MPE_E} = \pm\,[0.9\ \mu\mathrm{m} + (L/350)]\ \mu\mathrm{m} \tag{8.3}$$

$$\mathrm{MPE_P} = \pm\,1.00\ \mu\mathrm{m} \tag{8.4}$$

$$\mathrm{MPE_{Tij}} = \pm\,1.90\ \mu\mathrm{m}, \tag{8.5}$$

where $\mathrm{MPE_E}$ is the maximum permissible machine error, L is the measured length in mm, $\mathrm{MPE_P}$ is the maximum permissible probing error, and MPE_{Tij} is the maximum permissible tangential scanning probing error.

8.2.5   CMM Measurement Strategy

CMM machine consists essentially of a probe supported on three mutually perpendicular (X, Y, and Z) axes. Procedure for simple measurements on a CMM includes

(1) calibration of the probe system;
(2) defining datum(s) on the work piece;
(3) performing measurement(s);
(4) computing the required dimensions from measurements made in Step 3;
(5) assessing conformance to specifications.

The test room environmental conditions have to be adjusted to the standard specifications [15]. Mainly, room temperature of 20 ± 0.5°C and relative humidity of 50 ± 5% should be ensured around the CMM machine for the sake of verification. In addition, the probe specifications and the operating conditions should also be specified. Table 8.2 includes the specifications of the CMM setup measurements strategy.

Table 8.2   CMM setup measurements strategy

CMM strategy parameters	Specifications
Master probe radius	4.0000 mm
Reference Sphere radius	14.9942 mm, S = 0.0001 mm
Used long probe radius	4.0000 mm, S = 0.0001 mm
Machine traveling speed	20 mm/s
Probe scanning speed	15 mm/s
Fitting method	LSQ technique
No. of ROUNDNESS scanning points	136 points
No. of STRAIGHTNESS scanning points	32 points, step width = 3

8.3   Experimental

An experimental measurement program has been designed and implemented throughout this investigation in order to demonstrate how to apply the precision metrology in inspecting the crucial replaceable engine parts during an overhaul process. Cylinder block and cylinder head crucial surfaces which dominantly affect the whole engine reliability and durability have been inspected. In addition, engine performance parameters such as compression pressure attained in cylinders have been measured and correlated to the dimensional and geometrical measurements obtained. The details of this experimental program are presented in the following sections.

8.3.1 Cylinder Block Measurements

The setup of the CMM machine table during the inspection process of the engine cylinder block attributes is illustrated in Fig. 8.6. Precision multi-coordinate measurements of the cylinder bore diameter and roundness at 12 different transverse sections along the traveling axial distance of the piston, together with the longitudinal straightness along 12 equispaced paths, have been implemented. In other words, dimensional and geometrical (form and location) measurements have been carried out with the objective of error determination.

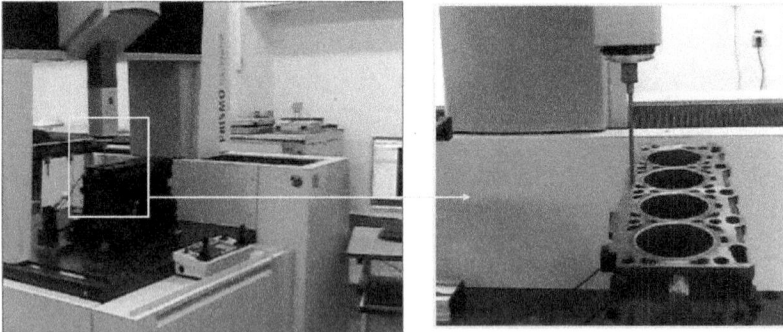

Figure 8.6 Photo of CMM setup arrangement for engine cylinder block measurements.

8.3.2 Cylinder Head Measurements

Figure 8.7 depicts the CMM measurement setup for the engine cylinder head inspection process carried out on the lapped portion of valves seats tapered surfaces. Figure 8.8 shows schematically the locations of the measured dimensional parameters (d_1, d_2, d_3, and d_4) from which sample results have been tabulated in Table 8.3.

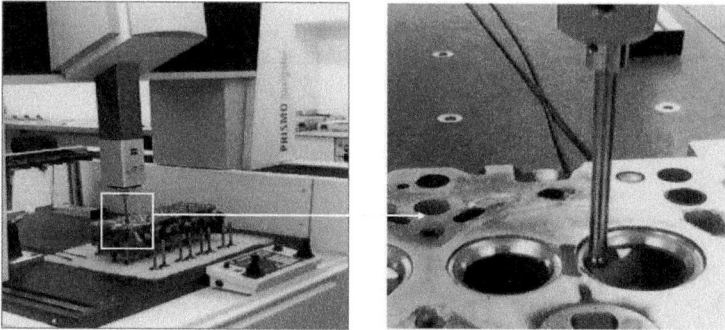

Figure 8.7   Photo of the CMM setup arrangement for the cylinder head inspections.

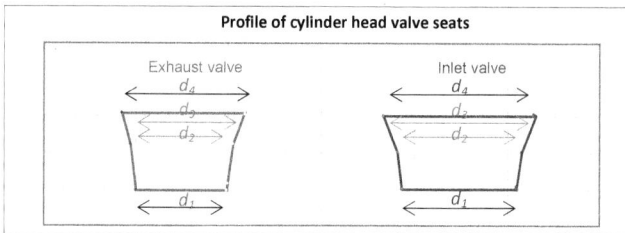

Figure 8.8   Schematic of the valves seats measurements locations.

Table 8.3   Sample results of dimensional measurements on the valve seat surface

Cylinder no. 1							
Exhaust valve seat				**Inlet valve seat**			
d_1	d_2	d_3	d_4	d_1	d_2	d_3	d_4
28.7	31.3	33.9	37.5	33.8	36.9	40.9	41.5

Furthermore, the common secured tight lapped area of contact between the valve and seat could be trigonometrically computed as aforementioned for cross-checking with the pixel size technique results. In addition, the roundness of the lapped area at any place on the valve seat could also be inspected wherever needed. Sample results of such measurements could be seen in Fig. 8.9. Also, the angle of inclination of the valve seats could be measured of which sample results are tabulated in Table 8.4.

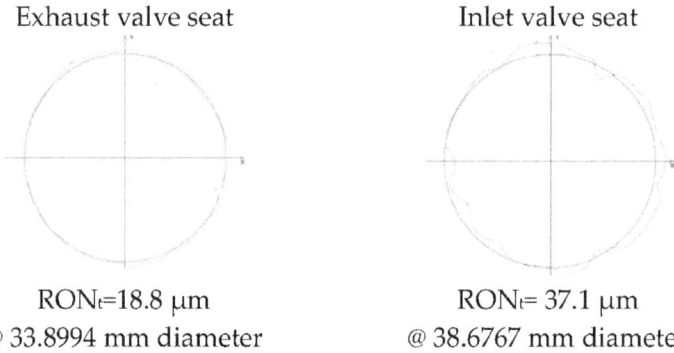

Exhaust valve seat

RON_t=18.8 µm
@ 33.8994 mm diameter

Inlet valve seat

RON_t= 37.1 µm
@ 38.6767 mm diameter

Figure 8.9 Sample results of roundness measurements on valve seat lapped portion.

Table 8.4 Sample results of measured angles of the valves seats surfaces

Engine cylinders	Cylinder no. 1		Cylinder no. 2		Cylinder no. 3		Cylinder no. 4	
Valves	Ex.	Inlet	Ex.	Inlet	Ex.	Inlet	Ex.	Inlet
Angle, °	44	54	53	42	53	52	49	47

8.3.3 Valve Measurements

Thus the contact lapped area between each valve and its seat could be precisely determined by applying the trigonometric relation of Eq. (8.2) and substituting the measured values of a, d_2, and d_3 for both inlet and exhaust valves of cylinder no. 1 as an example. Values of the resulted lapped areas are presented in Table 8.5.

Table 8.5 Calculated lapped areas of valves

Engine cylinder	Cylinder no. 1	
Valves	Ex.	Inlet
Area, mm^2	195.4	236.4

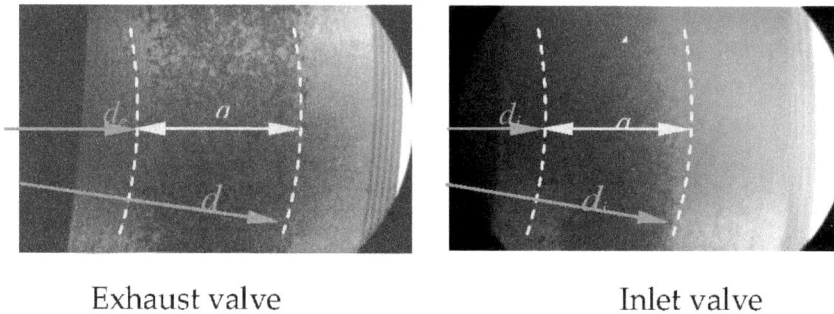

Exhaust valve   Inlet valve

Figure 8.10 Sample camera images of lapped areas on a valve surface.

8.3.4 Piston and Ring Measurements

Figure 8.11 illustrates schematically relative circular locations of a piston ring when it is freely opened and when it is restrained in the piston groove by the bore surface of the cylinder liner. The distance s is defined as the gap when the ring is uncompressed, while the gap s_1 is referred to as the closed gap or end clearance which is the minimum gap remained when the ring is installed in the cylinder bore. During operation, the ring slides up and down, touching the bore wall. If the cylinder bore is not completely round and straight, the ring gap repeatedly opens and closes. The resulting stresses are likely to break the ring [9]. A lack of lubrication also causes material failure. In four-stroke engines, the top (compression) ring is used mainly for sealing combustion gas with the help of the second one. The other bottom couple of rings are mainly to scrape and control the lubricating oil [9]. In this context the inspection program demonstrated some measurements carried out on a piston and top compression ring using both CMM machine together with a digital vernier. The results are tabulated in Table 8.6.

Table 8.6 Piston and ring measurements

Piston nominal diameter, d_p	85.40 mm
Ring nominal free diameter, d_f	90.05 mm
Ring nominal solid diameter, d_s	84.60 mm
Cylinder nominal bore diameter, d_c	86.00 mm

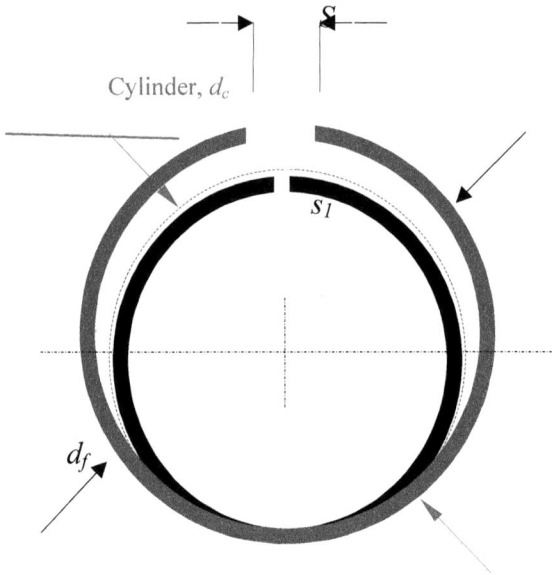

Figure 8.11 Nomenclature of a piston ring at open and closed conditions.

8.3.5 Measurement of Engine Performance

Engine manufacturing quality assurance together with proper operational conditions ensures reduced maintenance and maximum availability. The most reliable method of measuring cylinder pressure leak is to measure the compression pressure of each cylinder as it has become the standard method of determining the condition of an engine. An analog cylinder pressure dial indicator (Model G-320HD) has been adopted to measure such compression pressure after subjecting it to a calibration process according to DKD R 6-1 specifications. The pressure indicator calibration curve is shown in Fig. 8.12.

8.4 Uncertainty in Measurements

Geometrical form measured deviations of roundness (RON_t) and straightness (STR_t) have been extra analyzed by applying the uncertainty analysis concept according to ISO *Guide* (GUM) [17, 18]. Equations (8.6–8.9) have been applied on the measured values of roundness and straightness of cylinder bore surfaces:

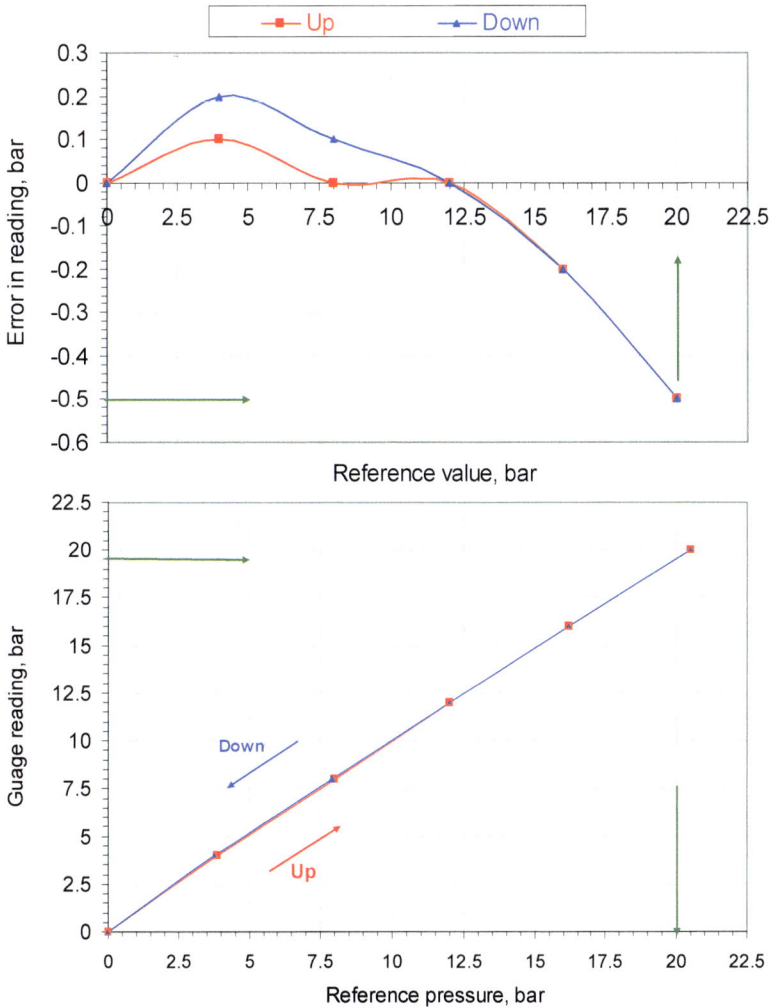

Figure 8.12   Calibration of pressure tester gauge.

$$M_{AV} = \frac{1}{n} \sum_{i=1}^{n} x_i \tag{8.6}$$

$$SD = \sqrt{\frac{1}{n} \sum_{i=1}^{n} (x_i - M_{AV})^2} \tag{8.7}$$

$$U_C = \frac{SD}{\sqrt{n}} \tag{8.8}$$

$$U_A = K \cdot U_C, \tag{8.9}$$

where x_i is the measured value of roundness or straightness in µm, n is the number of repeated test samples for target measurements, M_{AV} is the mean average value, and SD is the standard deviation. While U_C is the combined standard uncertainty due to measurement repeatability and/or statistical analysis, and U_A the expanded uncertainty (type A), where K is the standardized variable that is called coverage factor which equals 1.96 according to a confidence level of 95% [6]. It worth mentioning that type B source of uncertainty is not accounted for by U_A values because of its relative insignificance.

Figure 8.13 illustrates the expanded uncertainty U_A in 240 separated tests of dimensional measurements of engine cylinder bores. Figure 8.14 illustrates the expanded uncertainty U_A in 240 separated tests of geometrical roundness form measurements of engine cylinders. Figure 8.15 illustrates the expanded uncertainty in 240 separated tests of geometrical straightness form measurements of engine four-cylinder walls. The accuracy and uncertainty of these measurements have been determined and found to be within the acceptable standard suitable limits.

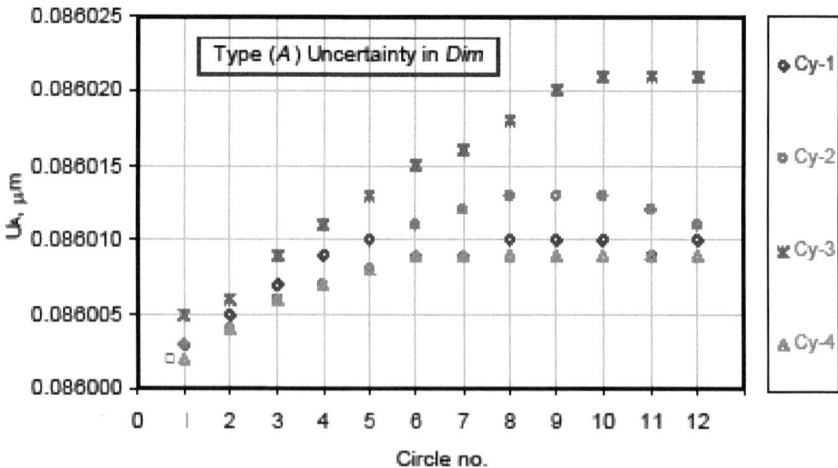

Figure 8.13 Uncertainty in dimensional deviation measurements of engine liners.

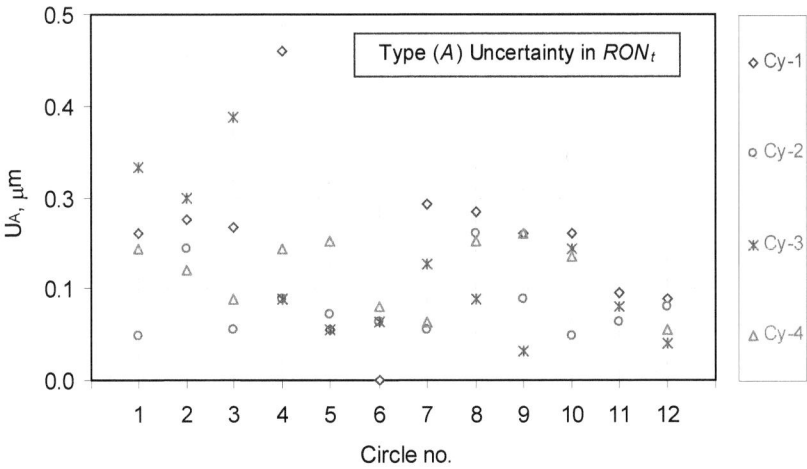

Figure 8.14   Uncertainty in roundness deviation measurements at 12 circles along engine liners.

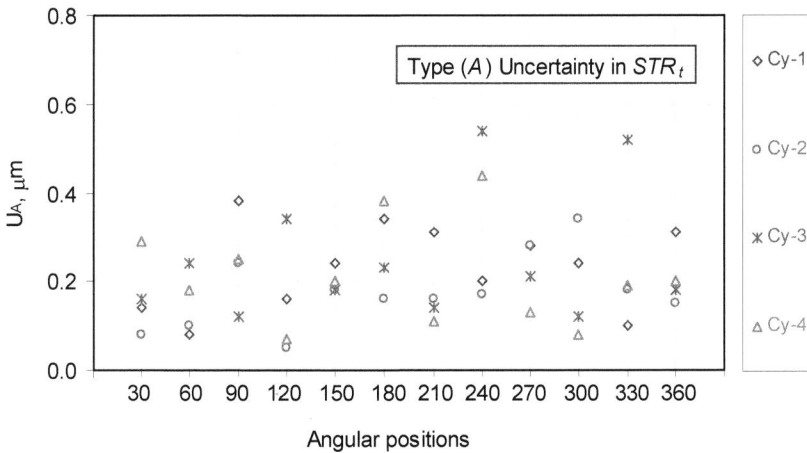

Figure 8.15   Uncertainty in straightness deviation measurements along cylinder liners.

8.5   Results and Discussion of Engine Inspection

The dimensional and geometrical form and location deviations of the cylinder liners fitted in the engine block under investigation in addition to the results of the compression pressure attained in the cylinders have been graphically presented in Figs. 8.16–8.21. The results have been discussed and interpreted in the following sections.

8.5.1 Results of Dimensional Deviations

The deviations in the bore diameter of the four cylinder liners of the test engine under investigation at 12 locations along the effective traveling stroke have been displayed in Fig. 8.16 as referred to a nominal bore diameter of 86.00 mm. Each point on the curve represents the average mean of five similar repeated measurement results. The 12 diameters of circles from no. 1 on the side of the top dead center (TDC) down to no. 12 near the bottom dead center (BDC) for the four cylinders have been precisely measured. The deviation increases as the measured circle approaches the BDC with the largest deviation detected in cylinder no. 3. This may be attributed to the lack of quality control and the associated metrology equipment available during manufacturing and/or finishing processes.

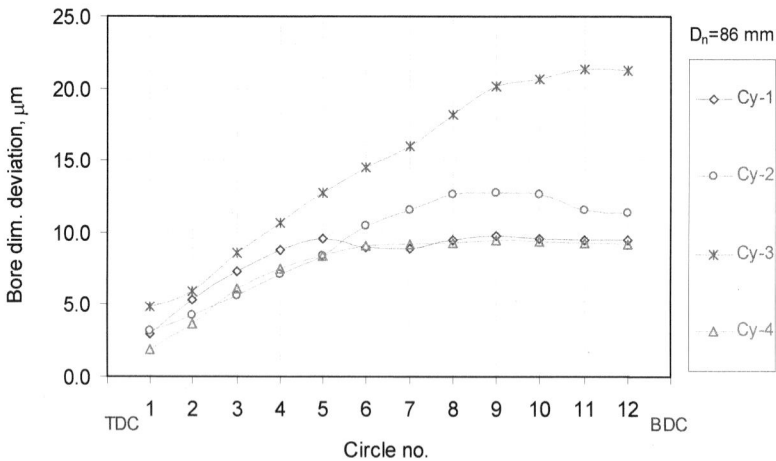

Figure 8.16 Deviation in cylinder bore diameters at 12 transverse locations.

8.5.2 Results of Form Deviations

The results of roundness deviation (RON_t) and straightness deviation (STR_t) as geometrical form deviation as assessed developed inspection program carried out on the cylinder block are presented in Figs. 8.17, 8.18, respectively. It can be generally observed that the RON_t increases as the measurement location approaches the BDC location reaching a maximum value of 19.52 μm in cylinder no. 3 with a 67.93% increase. Whilst, maximum values of RON_t have been found to be 16.24, 11.40, and 9.84 μm with 57.88%, 56.32%, and 53.25% increase in cylinders no. 4, 1, and 2, respectively. It can be observed that cylinder no. 3 is consistently the worst regarding both the

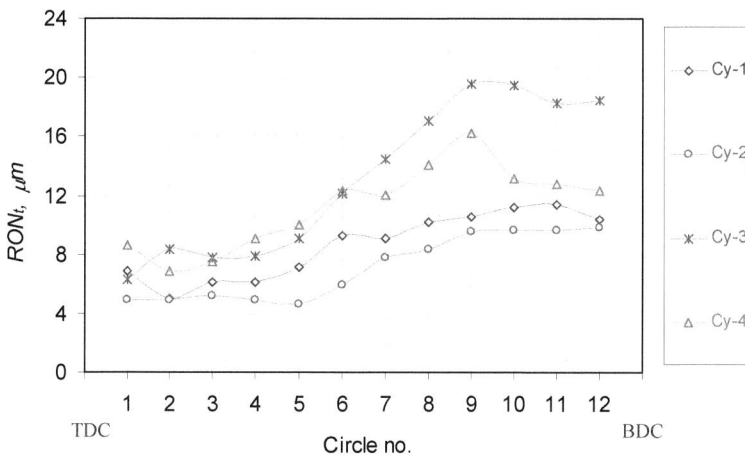

Figure 8.17 Roundness deviation at 12 transverse locations.

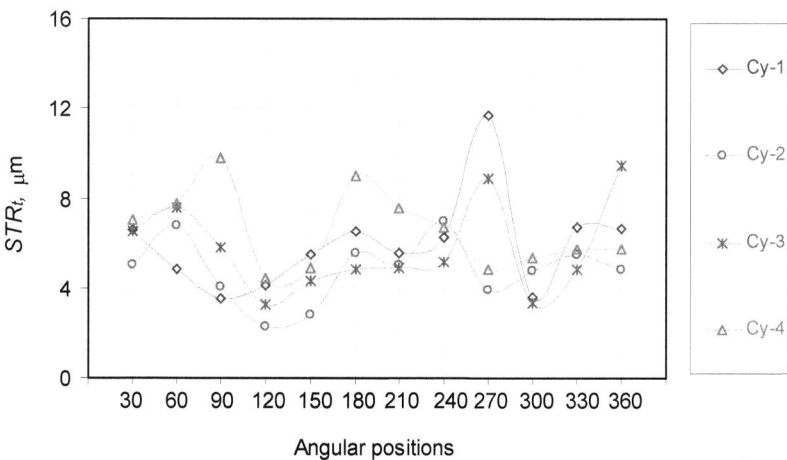

Figure 8.18 Straightness deviation at 12 longitudinal locations.

dimensional and the geometrical form deviations presented in Figs. 8.16 and 8.17. Figure 8.18 discloses the average straightness deviation magnitudes along 12 equispaced tracks on the interior surface of the cylinder bores. The wavy pattern of the curves within a domain lying between 3 and 12 µm indicates the quality of geometrical form control. The deviation of nine microns shown in Fig. 8.18 may be attributed to the deviations in the guides of the machine tool.

8.5.3   Results of Location Deviations

Figure 8.19 depicts the deviations in the polar coordinates r and θ of the centers locations of the 12 measured transverse circles described in Figs. 8.3 and 8.4. These deviations may be assumed to describe the coaxiality of each cylinder referred to the CMM setup system of reference. Figure 8.19a discloses the deviation of the polar radius r along the axes of the four cylinders with respect to an arbitrary nominal value selected for each cylinder. In addition, Fig. 8.19b describes the deviations of the θ polar coordinates of the same 12 measured centers for the four cylinders. Based on these assumptions, it can be concluded that the total amount of out of coaxiality in cylinder no. 1 showed comparatively a minimum value of 22.80 µm, while cylinder no. 4 experienced maximum total out of coaxiality of 41.86 µm.

8.5.4   Results of Engine Compression Pressure

The instantaneous peak compression pressure attained in cylinders has been adopted to represent the engine performance. It reflects the degree of internal sealing and tightness between the piston with its rings and the cylinder liner and between the valves and their seats as aforementioned. In order to correlate the conducted dimensional and geometrical inspection measurements in this chapter to the engine performance, Fig. 8.20 presents a bar chart describing the attained peak compression pressure in the four cylinders. It can be seen that cylinder no. 3 produces minimum compression pressure, which means that it suffers from power and pressure leak due to slackness between the contacting surfaces. This validates and interprets the precise and accurate dimensional and geometrical inspection results presented in Figs. 8.16, 8.17, and 8.19.

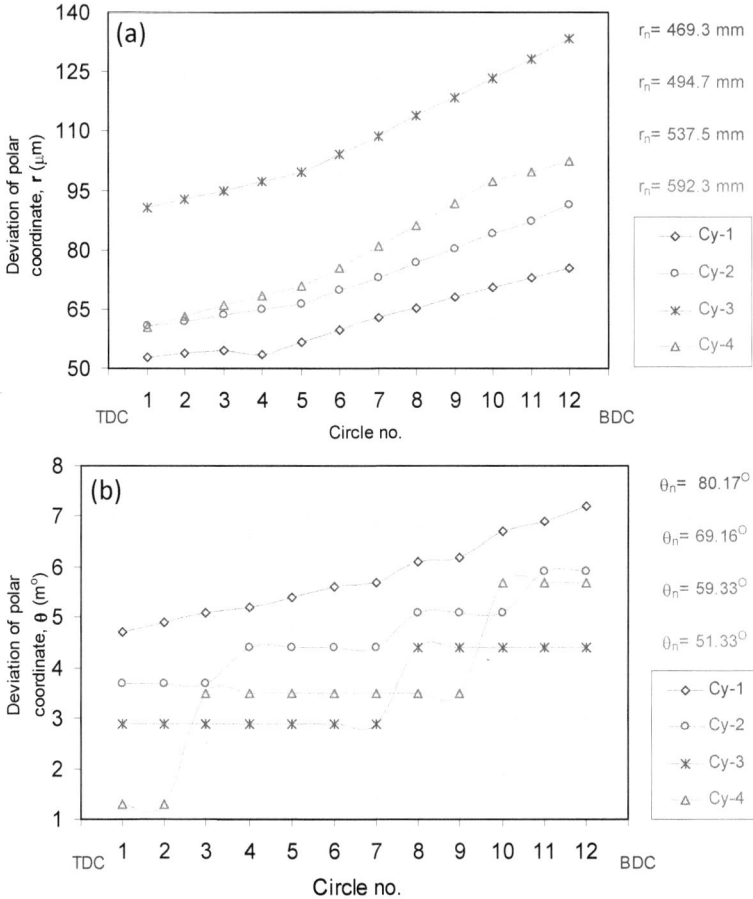

Figure 8.19 Relative deviations in location of the 12 centers polar coordinates. (a) Mean value deviation of polar coordinate *r*. (b) Mean value deviation of polar coordinate *θ*.

For these reasons, many applications using CMM integrals in automotive engine manufacturing processes can gain instant and reliable information of measurement on the production line (see Fig. 8.21). Figure 8.22 shows the rule image of the valve set for cylinder head of the engine [19].

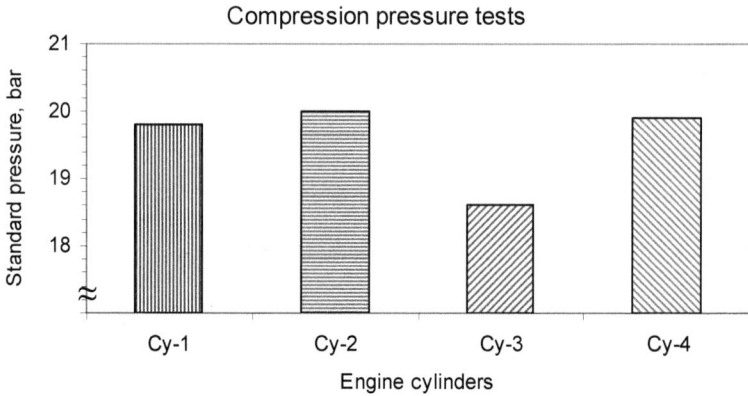

Figure 8.20   Engine cylinders peak compression pressure.

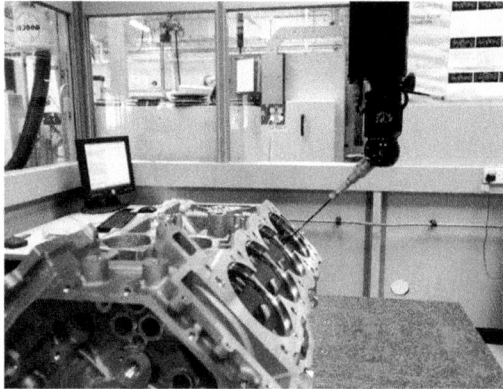

Figure 8.21   Quality control room of engine block measurement.

Figure 8.22   Valve set of cylinder head for engine block.

8.6   Conclusion

The results of implementing a developed dimensional and geometrical deviation inspection program on replaceable engine parts during an overhaul have been presented, discussed, and interpreted. The following conclusions can be drawn:

(1) A comprehensive inspection program has been developed to exploit the advanced precise dimensional and geometrical measurements in automotive engine repair and overhaul.

(2) Crucial engine parts that dominantly affect the engine performance such as cylinder liners, pistons and rings, and valves and seats can be successful inspected.

(3) The correlation between the instantaneous peak pressure and the departure from the nominal diameter and roundness of cylinder bore has been demonstrated.

(4) Inspection programs using precision dimensional measurements carried out on engine parts were successful in predicting the engine performance.

Acknowledgments

The author would like to thank Dr. Alaaeldin Eltawil of the Pressure Metrology Lab at NIS for his kind cooperation. The author appreciates the contributions of Prof. Dr. M. K. Bedewy, Cairo University (Allah place him) and Prof. Dr. S. Z. Zahwi, NIS, and their useful comments during final revision of the work.

References

1. S. Feng, A Dimensional Inspection Planning Activity Model, *Journal of Engineering Design and Automation: Special issue on Tolerance and Metrology for Precision Manufacturing*, vol. 2, no. 4, pp. 253–267, 1996.

2. R. R. Inman, D. E. Blumenfeld, N. Huang, and J. S. Li, Designing Production Systems for Quality: Research Opportunities from an Automotive Industry Perspective, *International Journal of Production Research*, vol. 41, no. 9, pp. 1953–1971, 2003.

3. D. E. Whitney, The Role of Key Characteristics in the Design of Mechanical Assemblies, MIT, Cambridge, Massachusetts, ISSN 0144-5154, pp. 315–322, USA, 26/4, 2006.

4. K. H. Chung, Application of Augmented Reality to Dimensional And Geometric Inspection, PhD thesis, Virginia Polytechnic Institute, USA, 2002.

5. B. Kilic, J. A. Aguirre-Cruz, and S. Raman, Inspection of the Cylindrical Surface Feature After Turning Using Coordinate Metrology, *International Journal of Machine Tools & Manufacture*, 47 pp. 1893–1903, 2007.

6. S. H. R. Ali, H. H. Mohamed, and M. K. Bedewy, μ-Scale CMM: A Proposed Diagnostic Tool Based on Evolved Deviations in Geometrical Measurements, *Metrology and Measurement Systems, Warsaw, Poland*, vol. XV, no. 3, pp. 363–375, 2008.

7. T. K. Garrett, K. Newton, and W. Steeds, The Motor Vehicle, Thirteenth Edition, Reed Educational And Professional Publishing ltd, ISBN 07506 4449 4, 2001.

8. R. Lewis and R. S. Dwyer-Joyce, *Automotive Engine Valve Recession*, Series Editor, Duncan Dowson, ISBN 1 86058 358 X, 2002.

9. H. Yamagata, *The Science and Technology of Materials in Automotive Engines*, Replika Press Pvt Ltd, India, ISBN-10:1-85573-742-6, First Pub., 2005.

10. A. Rossi, A Form of Deviation-Based Method for Coordinate Measuring Machine Sampling Optimization in an Assessment of Roundness, *Journal of Engineering Manufacture*, vol. 215, no. 11, pp. 1505–1518, 2001.

11. B. Muñoz, M. Ramírez, V. Díaz, Development of a New Methodology for Vehicle Steering System Inspection, *Journal of Automobile Engineering*, vol. 220, no. 11/2006, pp. 1515–1526.

12. J. A. Calvo, V. Díaz, E. Olmeda, J. L. San Román, A. Gauchía, Procedure to Verify the Maximum Speed of Automatic Transmission Mopeds in Periodic Motor Vehicle Inspections, *Journal of Automobile Engineering*, vol. 222, no. 9/2008, pp. 1615–1623.

13. T. Hassan, Applications of Computer Vision in Metrology, MSc thesis, Ain Shams University, Cairo, Egypt, 2006.

14. T. Tsai, Camera Calibration for a Manufacturing Inspection Workstation, Intelligent Systems Division, National Institute of Standards and Technology, Gaithersburg, MD 20899–8230, January 2005.

15. International Standard: Geometrical Product Specifications (GPS)-Acceptance and reverification tests for Coordinate Measuring Machines (CMM)-Part 1: CMMs used for Measuring Size, ISO 10360-2, 15-12-(2001).

16. International Standard: Geometrical Product Specifications (GPS)-Acceptance and Reverification tests for Coordinate Measuring Machines (CMM)-Part 2: CMMs used for Measuring Size, ISO 10360-2, 15-12-(2001).

17. Guide to the Expression of Uncertainty in Measurement, International Organization for Standardization, Geneva, Switzerland, 1995.

18. K. Birch, Measurement Good Practice No. 36: Estimating Uncertainties in Testing, An Intermediate Guide to Estimating and Reporting Uncertainty of Measurement in Testing, British Measurement and Testing Association, NPL, UK, March 2003.

19. Precision Manufacturing, Cosworth Co, UK. Website: http://www.cosworth.com/solutions/precision-manufacturing/capabilities/.

Chapter 9

Metrology as an Identification Tool for Worn-Out Air-Cooled Diesel Engine

9.1 Introduction

This chapter aims at exploiting the accurate precise measurements of the CMM machine in exploring and investigating the wear happening between the solid surfaces in contact. Geometrical and dimensional measurements using precision measuring devices are crucial during the manufacturing processes of parts to ensure their compliance with the design requirements [1]. In addition, those accurate measurements may also be employed with reference to their benchmark values to monitor the extent and severity of functional deterioration of the parts, especially those working with their surfaces during service. This helps the maintenance engineer make proper decisions regarding their forthcoming maintenance plan and/or repair actions. Thus, the durability and reliability of the parts and the assembly would be favorably affected.

Air-cooled diesel engines, for instance, are commonly used in heavy-duty transport fleet applications due to their high performance, efficiency, and low fuel consumption. The surface contact problems between cylinders and pistons through their rings are vital to the engine performance within the adverse operating conditions of high pressure, temperature rise, and

Automotive Engine Metrology
Salah H. R. Ali
Copyright © 2017 Pan Stanford Publishing Pte. Ltd.
ISBN 978-981-4669-52-8 (Hardcover), 978-1-315-36484-1 (eBook)
www.panstanford.com

high relative velocity of the contacting surfaces [2–4]. Fine finish and surface treatment together with proper geometrical and dimensional tolerances standards implementation are required in order to ensure good sealing between the cylinder wall and the piston rings, good load carrying capacity, good lubrication conditions, less friction, suitable wear resistance, low translated vibration levels, high engine efficiency, and longer service life span [5, 6]. The main function of the piston rings assembly is to provide a good dynamic sealing between the combustion chamber and the crankcase during compression and power strokes. Reasonable sealing minimizes power loss due to charge escape from the combustion chamber within suitable ring expansion gap and limited friction force. For long sealing service life, friction and wear between piston rings and the cylinder wall have to be properly controlled [7, 8]. They are controlled by the lubrication of the interface with dry lubrication of cylinder bore material composition besides an oil film thick enough to separate the asperities of piston rings and the cylinder surface [3–7]. The friction loss varies according to the piston velocity between the top dead center (TDC) and the bottom dead center (BDC), where the oil film thickness depends on the instantaneous relative velocity of the piston ring, which varies from zero at TDC and BDC to the maximum in the middle section. This means that wear conditions will vary along the piston ring traveling distance, from mild to severe [9].

Normally the cylinder bore is not cylindrical along its entire length. Practically, the bore distortion causes loss of conformity between piston rings and cylinder wall, which in turn causes some problem for oil film distribution. Variation in the oil film thickness exposes piston rings and the cylinder to the whole spectrum of lubrication regimes, from mixed and probably elastohydrodynamic to full film hydrodynamic lubrication [5, 7, 10]. Consequently, different wear mechanisms will develop geometrical departures in transverse sections along the cylinder bore [9].

TDC location on the bore suffers heavily from oil starvation more than that at the BDC and its vicinity. Although the piston at both locations are kinematically characterized by marginal inversion velocity situations where it reaches zero before starting to get inverted, the most severe wear is expected to appear at the TDC due to the oil shortage, while at the BDC the oil is available either from the source or due to gravity. However, the BDC may also experience high wear rate due to the existence of hard grit and wear debris accumulated by the gravity at this location and the neighboring area. The middle location and the nearby zone, where the piston velocity reaches its maximum value, mild wear only is expected because the oil film becomes dynamically thick enough to separate the mating solid surfaces and prevent metal-to-metal Contact [11].

Although there are many new advanced inspection equipment such as CMM machines, of which their use is so far only monopolized to the manufacturing fields [12, 13], rare published research work yet exists on the use of such advanced CMM metrology utilities in the field of engine health monitoring through geometrical departure measurements and analysis. The characterization of engine cylinder bore geometry and dimension is a twofold problem. The first is related to the applied techniques and quality standards adopted during manufacturing inspection process. This concerns the prescribed surface design parameters such as dimensional and geometrical tolerances, and surface roughness. The second is related to processing such data with the purpose of monitoring the changes that happened to the surface geometry and dimensions during engine service life span. This would help in two aspects: The first is related to maintenance decisions, while the second is related to design modifications. Research work has been done on surfaces with Gaussian distribution roughness, but cylinder wall fine finished surface with specified geometrical features and properties participate simultaneously together to controlling the environment that critically affects the engine functional performance and life [14–17].

Although the specified surface parameters represent advanced features, their definition is generally unrelated to any physical or mathematical properties of the surface topography [18]. The plotted accumulation of surface asperities heights according to the Gaussian distribution appears as straight-line scales. For transitional surface topography, such a scale appears as two intersecting straight lines. The slopes of the lines are proportional to the standard deviations of the two distributions, while the point of intersection represents the depth of transition from one finish to another. Difficulties encountered using this technique to apply, has recently solved with developing advanced calculations software [12].

On the other hand, the numerical description of the changes in the operating surface geometry during service life span necessitates detecting and follow up the surface geometrical deviations. However, some changes occur in such a way that a band of surface fine wavelength may disappear. Hence, Fourier transformation analysis is needed in this case to determine the surface power spectrum using special software to characterize the changes in the surface straightness and roundness relevant to operation environment changes [11]. Statistical calculation analysis of combined standard uncertainty (type A) is also needed for CMM measurements [20].

The purpose of this work is to demonstrate the employment of accurate precise surface geometrical and dimensional measurements to monitor and follow up the extent of severity of wear changes in a worn-out cylinder of an automotive diesel engine as related to the resulted geometrical distortions in both transverse directions (out-of-roundness, and derived concentricity), and longitudinal directions (out-of-straightness). Thus, design improvements and/or correction actions to the scheduled maintenance plan could be suggested in the light of the analysis of the obtained measurements within the relevant uncertainties. Innovative design modification and inspired ideas may also be pointed at for the sake of extending the engine service life span and minimizing the running operational and maintenance expenses.

9.2 Cylinder Forces and Surface Measurements

9.2.1 Dynamic Friction Force

Gas pressure due to combustion represents the essential axial force acting on the piston crown area to move it downwards against reciprocating mass inertia. F_n is the instantaneous sum of the normal acting forces on piston pin (Fig. 9.1). Reciprocating piston motion on angular movable connecting rod generates a variable piston side force F_s. An axial transmitted force F_a of the crankshaft due to clutch engagement force and gear force components affects the cylinder wall. The resultant of piston forces F_s and F_a attacks the wall at an angle with F_s. The angle value varies as a function of the force amplitude to generate a resultant force causing rotation around the cylinder axis. Dynamic friction force F_f has been produced due to relative motion of piston rings with respect to the cylinder wall under the effect of the resultant force in a spiral like motion. This causes the cylinder bore to wear at rates corresponding to the resultant force amplitude and direction to generate eventually a cylinder out-of-roundness (OOR) and out-of-straightness (OOS).

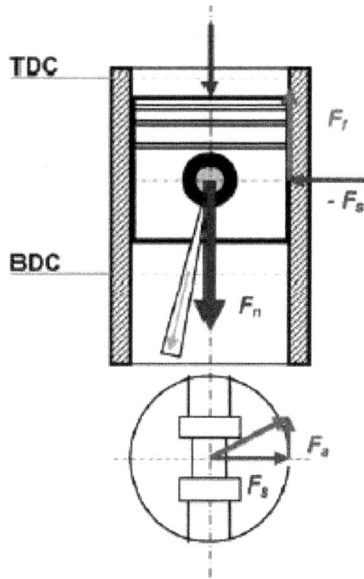

Figure 9.1   Forces acting on the cylinder bore.

9.2.2   Surface Geometry Measurements

Geometrical and dimensional characteristics of the cylinder bore surface have been measured using a computerized coordinate measuring machine (CMM) equipped with a contact-scanning probe and a least square (LSQ) computing algorithm. The CMM used throughout this work was Carl Ziess bridge model available at the Engineering and Surface Metrology Lab, Precision Engineering Division, National Institute for Standards (NIS) at Egypt. It is capable of producing accurate results with a reasonable repeatability and reproducibility for the surface geometrical departure features. The maximum permissible specific error value of the used CMM machine can be judged using the following equation:

$$\text{MPEE} = (0.9 \ \mu m + (L/350)) \ \mu m, \tag{9.1}$$

where L is the measured length in millimeters.

The CMM measurement performance was verified according to ISO-10360.19 An experimental investigation has been conducted on an air-cooled diesel engine cylinder made of high quality gray cast iron (GG 25) having initially a design diameter of 110 mm and configuration shown in Fig. 9.2a. The chemical analysis and mechanical properties of the cylinder material are presented in Table 9.1, where HB is the Brinell hardness and σ_t is the tensile strength. The piston stroke is 140 mm. A straight Stylus tungsten carbide shaft probe with a ruby tip attached to PRISMO CMM machine was used to quantify the surface geometric and dimension departure characteristics of the cylinder bore. The CMM traveling speed was 40 mm/s and the probe scanning speed was 10 mm/s during measurements. The straightness measurements were carried out along four longitudinal equispaced locations, 90° apart around the circumference, at 1, 2, 3, and 4 as indicated in Fig. 9.2b. Cylinder bore roundness quantification was conducted at Sections I, II, III, and IV nearby TDC, midway, and BDC planes as shown in Fig. 9.2b. The surface geometrical and dimensional features were represented by mean average values of five repeated test measurements.

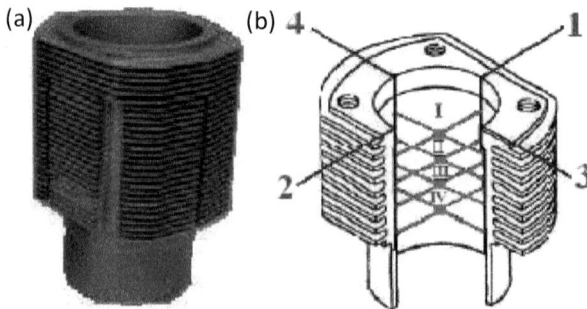

Figure 9.2 Engine cylinder configuration and locations of measurements. (a) Cylinder configuration. (b) Locations for roundness and straightness measurements.

Table 9.1 Cylinder material specifications

Chemical analysis, wt. %							Mechanical properties	
C	Si	Mn	P	S_{Max}	Cr	Ni	HB	$\sigma_{t, MPA}$
3.10	2.10	0.65	0.30	0.10	0.20	0.32	220	Min. 220

9.3 Uncertainty Assessment of Measurements

Mean average values and combined uncertainty of roundness and straightness measurements for the engine cylinder bore have been presented in Table 9.2, where MAV is the mean value of five repeated test measurements, SD is the standard deviation, and UC is the combined standard uncertainty due to measurement repeatability ($UC = SD/\sqrt{n}$), where n is the number of repeated tests for each target measurement [19]. It worth mentioning that type B source of uncertainty is not accounted for by UC values because of its relative insignificance to the OOR and OOS. The accuracy and uncertainty of these measurements have been determined and found to be within the acceptable standard limits.

Table 9.2 Measurements of both roundness and straightness together with the uncertainty assessment

Positions \ Tests	Test 1	Test 2	Test 3	Test 4	Test 5	M_{AV}	S_D	U_C
1. Roundness, µm								
@ circle I	91.0	90.9	90.9	90.9	91.0	90.94	0.0548	0.0245
@ circle II	32.1	31.8	32.1	31.9	31.8	31.94	0.1152	0.0678
@ circle III	23.0	23.5	24.2	24.0	24.3	23.80	0.5431	0.2429
@ circle IV	18.2	18.3	18.3	18.5	18.9	18.44	0.2793	0.1249
2. Straightness, µm								
@line 1	71.0	71.0	70.3	70.5	70.1	70.58	0.4087	0.1828
@line 2	54.4	54.1	54.3	54.5	54.5	54.36	0.1673	0.0748
@line 3	13.6	13.8	13.7	13.6	13.7	13.68	0.0837	0.0347
@line 4	34.6	34.7	34.8	34.8	34.9	34.76	0.1140	0.0510

The roundness and straightness results of five repeated laboratory tests conducted on each one of the adopted four transverse Sections I, II, III, and IV, and the four longitudinal profiles 1, 2, 3, and 4, have been processed and presented in Figs. 9.3 and 9.4, respectively. The results are also tabulated in Table 9.2 and the calculated values of the relevant uncertainty are plotted in Fig. 9.5.

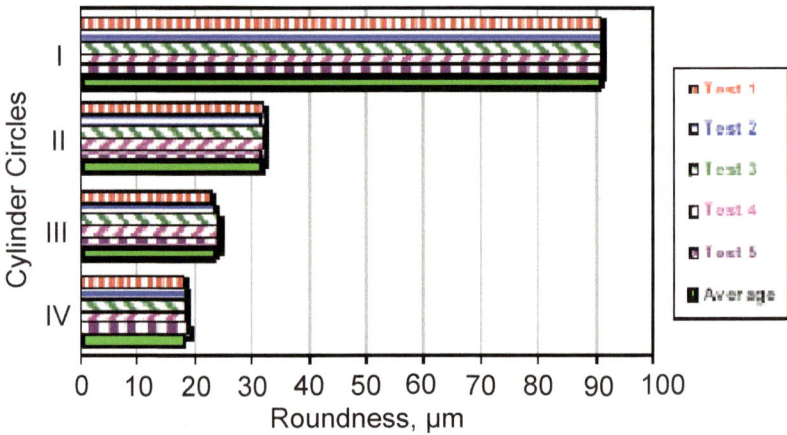

Figure 9.3 Bore roundness measurements at four transverse sections along piston stroke.

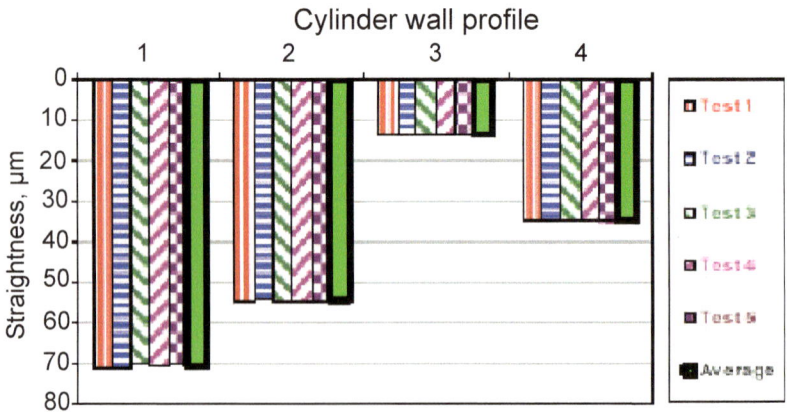

Figure 9.4 Bore straightness measurements at four longitudinal equi-spaced locations.

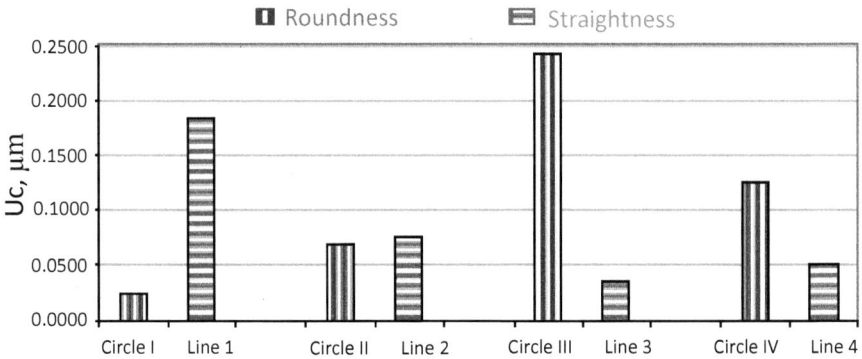

Figure 9.5 Uncertainty values of five repeatable tests of OOR and OOS.

9.4 Results and Discussion

Roundness, straightness, and concentricity averages for cleaned-up worn cylinder bore have been measured using accurate stylus surface scanning technique on a programmable CMM machine. The concentricity is represented by the relative roundness run out at the selected transverse Sections I, II, III, and IV with respect to circle I taken as a datum as shown in Fig. 9.2b. RMS averaged values of five similar arrays of measurements have been considered. The results have been presented, discussed, and interpreted.

9.4.1 Out-of-Roundness Measurement Results

Average out-of-roundness results (R_a values) have been processed for each measured circle on the bore surface and presented in Fig. 9.6 with reference to the nominal diameter, which is numerically computed and found equal 111 mm. The roundness is represented at each transverse section by the domain between the two virtual enveloping circles tangent to the distorted shape processed using LSQ fitting technique built in the machine as indicated in Fig. 9.6. Analysis of the roundness patterns of the cylinder bore, illustrated in Fig. 9.6, indicates the following points:

- The CMM machine software establishes a reference geometric feature of ideal regular form, deduced numerically from one or more realistic irregular scanned shapes. The established reference datum can be used in the assessment process of the run out values of the geometric features of the object under investigation.

- Circle I near the TDC, as expected due to lubricant starvation, depicted the highest average distorted dimension of DI = 111.1779 mm and the largest average out-of-roundness (R_a = 90.8 µm). Whereas, the smallest distorted dimension was depicted at section III (DIII = 111.0204 mm) in the vicinity of the mid-stroke point of the piston crown ring with average OOR value R_a = 23.8 µm, while the smallest out-of-roundness value was found nearby the BDC at circle IV (R_a = 18.4 µm).

- Roundness of circle I has two maximum amplitudes (arrow tips in Fig. 9.6) at points corresponding to location of resultant surface reaction $\sqrt{F_s^2 + F_a^2}$ of piston side force F_s and crankshaft axial force F_a. The side force amplitude and direction vary according to the nature of piston traveling displacement especially at compression and power strokes.

- Amplitudes of circle III have the smallest wear variation rate with relatively small out-of-roundness, which may be due to good lubrication conditions and light side forces at that location. However, roundness near the BDC (circle IV) has different directions of peak amplitude going with the indicated direction of cylinder distorted shape. This may be attributed to the stud clamping force situation (magnitude and direction) when the piston passes by this location. At both BDC and TDC marginal inversion locations, the loads on the piston generates a stringent translated piston dynamics.

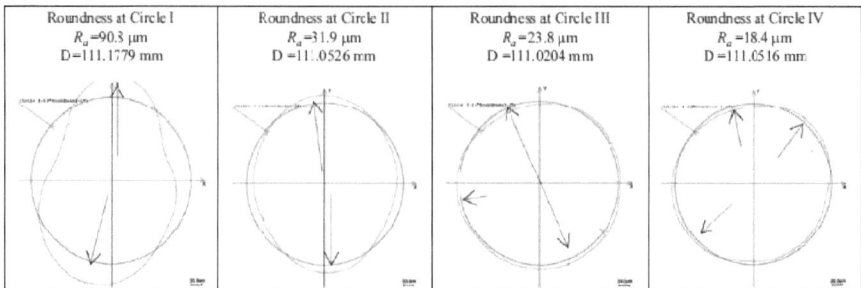

Roundness at Circle I	Roundness at Circle II	Roundness at Circle III	Roundness at Circle IV
R_a=90.3 µm	R_a=31.9 µm	R_a=23.8 µm	R_a=18.4 µm
D=111.1779 mm	D=111.0526 mm	D=111.0204 mm	D=111.0516 mm

Figure 9.6 Roundness sample measurement records of engine cylinder bore (maximum amplitudes).

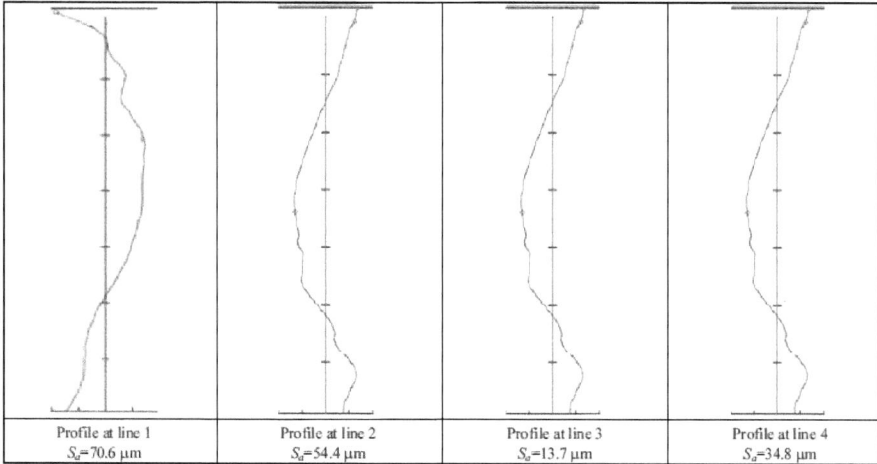

Profile at line 1	Profile at line 2	Profile at line 3	Profile at line 4
S_a=70.6 μm	S_a=54.4 μm	S_a=13.7 μm	S_a=34.8 μm

Figure 9.7 Longitudinal sample profiles of cylinder straightness (S_a is the averaged straightness of the profile).

9.4.2 Concentricity Measurements

Experimentally measured values of the relative roundness on the bore at different transverse locations (concentricity) have been found to be 39.1, 44.4, and 61.2 μm between circles I and II, circles I and III, and circles I and IV, respectively, with the axial center line of round profile I as reference datum as shown in Figs. 9.2 and 9.6. This would reflect the distortion resulting from the extremely severe wear mechanisms to which the engine cylinder bore was being experienced during service.

9.4.3 Out-of-Straightness Measurement Results

Figure 9.7 shows sample record of four averaged longitudinal profiles at equispaced locations 1, 2, 3, and 4 along the cylinder inner wall as indicated in Fig. 9.2 above. The maximum out-of-straightness value (S_a) processed from the measurements along each longitudinal profile represents the deviation domain around the relevant reference line obtained by applying LSQ fitting technique. Straightness profile sample records shown in Fig. 9.7 disclose the following points:

- Non-uniform wear rates are exposed along all averaged longitudinal profiles. It is clear that every point on the cylinder bore is subjected to different concurrent dynamic and environmental conditions of pressure, friction, lubrication scheme, sliding velocity, contact temperature, and contact force (orientation and magnitude). Thus, frequent evaluation of bore surface geometrical status is needed whenever possible to help monitoring the functional degradation and diagnosing the surface failure symptoms in anticipation, so that reasonable decisions can be made regarding surface treatment implementation and/or constructional design improvement inspiration.

- Maximum wear rates have been found to consistently lie within the TDC of the first pressure ring contact area for all averaged profile measurements of 70.6, 54.4, 13.8, and 34.8 µm. This may be explained by the bad tribological conditions at the TDC location as aforementioned. The largest value of straightness departure (70.6 µm) which lies on profile 1 was formed during power strokes as a direct response to large side force reaction at high combustion temperatures. These findings agree with a study carried out by Schneider et al. [5].

- Wear valleys of bore straightness have large values for the profile at points 1 and 2 of power and compression stroke ends (nearby BDC) due to side force reaction of concentrated piston inertia, while profile of points 4 and 3 have shown the smallest amplitude valleys, respectively.

- An extended valley of straightness has the first profile of power stroke until 70 mm long; it may be produced of piston-skirt stringent side pressure and combustion gas high temperature beside piston rotation around its pins under the effect of the friction force moment. Strong piston skirt dynamics accelerates the wear of the crankshaft axial movement control washers.

9.5 Conclusions

- Geometrical and dimensional micro-scale precision measurements of straightness, roundness, bore diameter, and concentricity of the internal surface of a worn-out engine cylinder have been executed using the CMM machine.

Compared to its original design GT&D tolerance limits, these measurements proved to represent a novel reliable diagnostic tool for the wear development and aggression monitoring. Scenarios of the probable adverse operating conditions during service may also be drawn. The dimensional measurements of the bore diameter at different transverse locations along the traveling stroke have assured previous findings using other different complicated measuring techniques. In turn, this may provide feedbacks to both the engine designer for modifications and the maintenance engineer for his forthcoming preventive and corrective maintenance plans layouts.

- Wear at the TDC and BDC transverse sections has been found to be much larger than the wear occurring at the middle of the stroke and near the BDC. This phenomenon is attributed to the continuous existence of lubricating oil film dynamically preserved at that location.

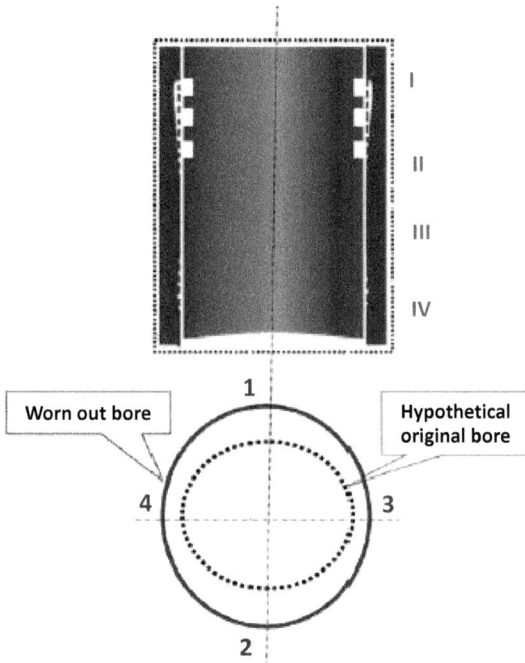

Figure 9.8   Illustration of geometric deviations in the cylinder liner.

- Precision CMM measurements may also provide an insight into the engine dynamics that may contribute to the excessive wear occurrence in the engine cylinder. The geometric deviation due to inhomogeneous wear has caused ovality in the bore where $S_{a1} > S_{a2} > S_{a4} > S_{a3}$, as depicted in Fig. 9.8. This may inspire the engine designer to introduce an innovative modification to the engine by developing a controllable cylinder-rotating device about its axis probably without having to dismantle the engine parts, so that the wear can be homogenized. Thus, the power loss due to friction and wear in the cylinder may be minimized and the engine operating life span may be rather prolonged. In addition, the maintenance expenses may be also reduced.

Acknowledgments

The author appreciates the contributions of Prof. Dr. M. K. Bedewy, Cairo University (Allah place him), and Prof. Dr. Hassan H. Dadoura, Helwan University, during the final revision of the work.

References

1. Y. H. Roh and C. M. Lee, A Study on Development of the High Precision Cam Measurement Apparatus and Analysis of Cam Manufacturing Error, *IJPEM*, vol. 26, no. 5, pp. 112–119, 2009.

2. N. W. Bolander, B. D. Steenwyk, K. Ashwin, and F. Sadeghi, Film Thickness and Friction Measurement of Piston Ring Cylinder Liner Contact with Corresponding Modeling Including Mixed Lubrication, *Conference of the ASME Internal Combustion Engine Division*, pp. 1–11, 2004.

3. M. J. Neale, *The Tribology Handbook*, 2nd ed., Butterworth-Heinemann, pp. D5.1–D5.5, 1995.

4. M. El-Sherbiny, *Cylinder Liner Wear*, 9th Leeds-Lion Symposium on Tribology, p. 132, 1982.

5. E. W. Schneider, D. H. Blossfeld, D. C. Lechman, R. F. Hill, R. F. Reising, and J. E. Brevick, Effect of Cylinder Bore Out-of-Roundness on Piston Ring Rotation and Engine Oil Consumption, SAE, Paper no. 930796, 1993.

6. Ohlsson, R., A Topographic Study of Functional Surfaces, PhD Thesis, Chalmers University of Technology, 1996.

7. P. Andersson and J. Tamminen, Piston Ring Tribology: A Literature Survey, Helsinki University, VTT Research Notes 2178, 2002.

8. N. H. Jayadas, K. P. Nair, and G. Ajithkumar, Tribological Evaluation of Coconut Oil as an Environment-Friendly Lubricant, *Tribology International*, vol. 40, issue 2, pp. 350–354, 2007.

9. J. Vatavuk and V. Demarchi, Improvement of Cylinder Liner Materials Wear Resistance, SAE, Paper no. 931671, 1993.

10. T. Nabnu, N. Ren, Y. Yasuda, D. Zhu, and Q. J. Wang, Micro-Textures in Concentrated Conformal-Contact Lubrication: Effects of Texture Bottom Shape and Surface Relative Motion, *Tribology Letters*, vol. 29, no. 3, pp. 241–252, 2008.

11. C. P. McNally, Development of Numerical Model of Piston Secondary Motion for Internal Combustion Engines, M.Sc. Thesis, Massachusetts Institute of Technology, 2000.

12. B. C. Jiang and S.-D. Chiu, Form Tolerance-Based Measurement Points Determination with CMM, *Journal of Intelligent Manufacturing*, vol. 13, no. 2, pp. 101–108, 2004.

13. B. Kilic, J. A. Aguirre-Cruz, and S. Raman, Inspection of the Cylindrical Surface Feature after Turning Using Coordinate Metrology, *International Journal of Machine Tools & Manufacture*, vol. 47, issues 12–13, pp. 1893–1903, 2007.

14. B.-G. Rosen, R. Ohlsson, and T. R. Thomas, Wear of Cylinder Bore Micro-Topography, *Wear*, vol. 198, issues 1–2, pp. 271–279, 1996.

15. A. Ramalho, A Geometrical Model to Predict the Wear Evolution of Coated Surfaces, *Wear*, vol. 264, issues 9–10, pp. 775–780, 2008.

16. C. Oner, H. Hazar, and M. Nursoy, Surface Properties of CrN Coated Engine Cylinders, *Materials and Design*, vol. 30, issue 3, pp. 914–920, 2009.

17. H. Hazar, Characterization of MoN Coatings for Pistons in a Diesel Engine, *Materials and Design*, vol. 31, issue 1, pp. 624–627, 2010.

18. A. L. George, and P. K. Nikolaos, Friction Model of a Marine Diesel Engine Piston Assembly, *Tribology International*, vol. 40, issues 10–12, pp. 1441–1453, 2007.

19. K. Birch, Measurement Good Practice No. 36–Estimating Uncertainties in Testing, *British Measurement and Testing Association*, 2003.

20. Geometrical Product Specifications (GPS), Acceptance and Reverification Tests for Coordinate Measuring Machines (CMM)—Part 2: CMMs Used for Measuring Linear Dimension, ISO 10360-2, 2001.

Chapter 10

Surface Metrology in Engine Quality

This chapter reviews the accurate precise measurements using surface metrology techniques in exploring and investigating the surface characteristics of engine materials in the range of micro- and nanometer scale. There are many parameters influencing the engine performance. Automotive engine cylinder, piston, and piston ring are fetal elements in the engine performance. Engine elements should be with high strength, hardness, heat-resisting and abrasion-proof property. Consequently, the investigation of the surface topography for the sensitive rubbing rotating elements of engine material and its quality affects the tribological response in the contact area between the cylinder wall and the piston ring [1]. Therefore, the inner surface of engine cylinder should be very fine and hard enough to resist the piston rings sliding friction in high temperature and pressure due to combustion [2]. So we can say that the performance of automotive engines certainly influenced by the used metrology techniques in the industry.

This chapter discusses the surface characterization of the cylinder wall, piston, and piston ring material using metrology techniques to understand the quality of surface performance. Moreover, the selection of the materials used for the internal combustion engine is also a very important parameter to be understood. Moreover, the current trend towards a compact engine with high power density and increased thermo-mechanical

Automotive Engine Metrology
Salah H. R. Ali
Copyright © 2017 Pan Stanford Publishing Pte. Ltd.
ISBN 978-981-4669-52-8 (Hardcover), 978-1-315-36484-1 (eBook)
www.panstanford.com

load increases the importance of tribological characteristics and requires new approaches in the area of cylinder working surfaces, which have been discussed in detail. Consequently, the cylinder inner surface is designed to withstand higher combustion pressure and higher piston speed. Furthermore, the precise matching of the cylinder bore with the pistons and piston rings leads to improved engine performance. There are conventional and new materials nowadays used in manufacturing internal combustion engine cylinders.

In order to reduce oil consumption and to ensure low friction and good wear resistance, the interaction of the cylinder and the piston ring (grey cast iron or steel) must be optimized [1]. Davis [2] pointed that most cylinder liners are made of gray cast iron. To increase their mechanical strength, addition of nickel, chromium, copper, and molybdenum is required. On the other hand, Nguyen [4] explained that as engine designs became more complicated, the weight of the engine and the vehicle had increased. Consequently, the desire among manufacturers to use lighter alloys that were as strong as cast irons arose. One such material used as a substitute was aluminum alloy. In the 1930s (due to problems with durability), aluminum alloy was used for the first time in engine blocks, and its use increased during the 1960s and 1970s as a way to increase fuel efficiency and performance; at that time the engine performance was not sophisticated. While, John Lenny Jr. in 2011 [5] presented a study of replacing the cast iron liners with aluminum for engine cylinder blocks. He believed that one approach to increasing an automotive fuel economy was by reducing vehicle weight and friction loss simultaneously. This is due to the increase of thermal efficiency of combustion. However, cast iron bores are needed in aluminum cylinder blocks because of their tribological characteristics. From his conclusions, hypereutectic Al–Si alloy for cylinder block becomes an increasingly attractive alternative for solving the tribological deficiency of most commercialized aluminum alloys.

As a development of metal matrix composite (MMC) in engine cylinder block, Manabu Fujine et al. [6] presented a new MMC which makes it possible to reduce engine dimensions and weight. Accordingly, the author achieves good wear properties and productivity, the MMC material for the cylinder block was developed under the following assumptions: Typical aluminum die-cast alloy containing 11% silicon and 2.5% copper is used for aluminum matrix, since conventional aspirating process is used to produce the preform. The preform consists of either fibers or fibers and particles. Stainless steel with nitride is used as the piston ring material. In parallel, Wilson Luiz Guesser [7] presented an overview of new trends in 2003 for diesel engines, opening a promising possibility of the use of compacted graphite iron (CGI).

Therefore, researchers and specialized companies are interested in the importance of coating operations for cylinder walls, piston, and piston ring. For many years, automotive companies have been funded scientific projects to improve quality, performance, efficiency and extended warranties. Consequently, in the past years coated operation has become very important are very important for corrosion resistance, wear resistance, lubricity, and uniform deposit thickness properties have been used to give great advantage in this automotive engines industry [8–9].

10.1   Engine Quality Using Metrology Techniques

The need for accurate dimensional measurements and quality engineering surfaces has become a necessary requirement to meet the challenges of modern technologies. The advanced soft metrology techniques include coordinate measuring machines, roundness testing facilities, surface roughness devices and optical methods used at micro- and nanometer scales [10]. Some studies pointed great development methods using optical and tactile techniques [10–23]. The interesting element of this device is a probe—magnetically fixed—which prevents damage (break). It is especially useful in the automotive industry, in the production of pumps and injectors, where without fear of probe damage CNC procedures to measure roughness, waviness, and outline in small holes could been used.

10.1.1   CMM Metrology Technique

Salah H. R. Ali [11] presented a research article on the performance investigation of CMM measurement quality using the Flick standard. He explained new experimental investigations to improve the measurement accuracy of dimension and roundness form of a coordinate measuring machine for particular measurement tasks. Measurements for transverse circle location of certified Flick standard have been carried out. The CMM output significant contributions from larger wave numbers than 25 UPR (undulations per revolution) are demonstrated at different scanning speeds and fitting algorithms. The experimental results show that the roundness error deviation can be evaluated effectively and exactly for CMM performance by using Flick standard. The measurement errors of corresponding different geometric evaluation algorithm and probe scanning speed (1, 2, 3, 4, and 5 mm/s) are obtained through repeated arrangement, comparison, and judgment. Ali concludes that the 2 mm/s probe speed gives smaller roundness error than 1, 3, 4, and 5 mm/s within 0.2: 0.3 μm.

ZEISS	Calypso 4.0.14	**Carl Zeiss**		Date	August 18, 2008
				Order	* order *
Part Number 2	CMM PRISMO	Drawing No. * drawingno *		Department: Operator Signature:	Dr. Salah
Measurement Plan Mazda Engine-A5				CAD View	

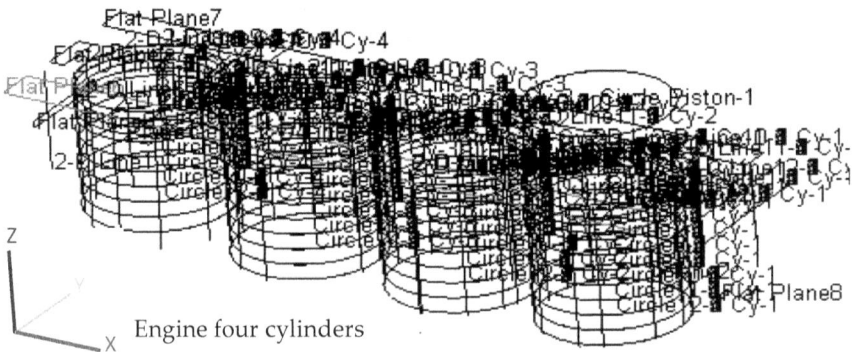

Figure 10.1 CMM result for Cylinder Block Measurement [12].

Salah H. R. Ali [12–13] developed an inspection program to be used as an advanced metrological device in conducting precise and accurate dimensional and geometrical measurements of automotive engines using the accurate precise measurements of CMM machine. Some measurement results of the CMM machine to inspect engine block attributes as shown in Fig. 10.1b. Ali proved the ability of this developed program to assess the real state for automotive engines in both new and used conditions accurately. He concludes that crucial engine elements that dominantly affect the engine performance, such as cylinder bores, piston rings, valves, and valve seats, can be reasonably inspected, and contact metrology techniques are good methods for surface feature characterizations.

10.1.2 AFM Metrology Technique

The atomic force microscope (AFM) as an advanced scanning probe microscope (SPM) is designed to measure local properties, such as height, friction, and magnetism using a sharp tip probe with 3–6 μm tall pyramid and 15–40 nm end radius. The SPM raster scans the probe over a small area of the sample to measure the local property simultaneously.

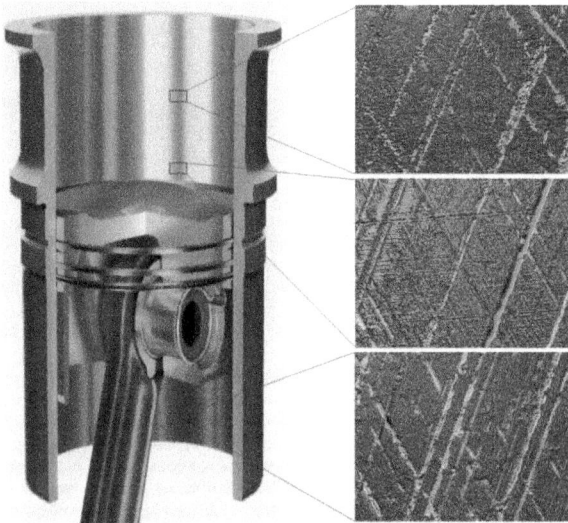

Figure 10.2 AFM result of cylinder surface at different positions.

The atomic force microscope is widely used in materials science and has found many applications in engineering nanotechnology. The AFM measures the surface characteristics and can measure the small depth of the produced structures exactly; an algorithm to engrave "invisible" small 3D structures in surfaces in closed loop control is possible [14]. The topography of engineering surfaces is commonly used to analyze surfaces before and after operation processes. The applications of texturing include pistons, brake discs, bearings, mechanical face seals, gas seals, hard disk sliders, machine tools guide ways, and other elements. The cylinder liner is made of much softer cast iron and becomes markedly different after having worn in the engine. The plateaux have been massively smoothed leaving the deep honing grooves clearly visible. Typical image results for the surfaces at different three positions as given in Fig. 10.2.

10.1.3 Scanning Electron Microscopy Technique

The scanning electron microscope (SEM) is a type of electron microscope that images a sample by scanning it with a beam of electrons in a raster scan pattern. The electrons interact with the atoms that make up the sample producing signals that contain information about the sample's surface topography, composition, and other properties such as conductivity.

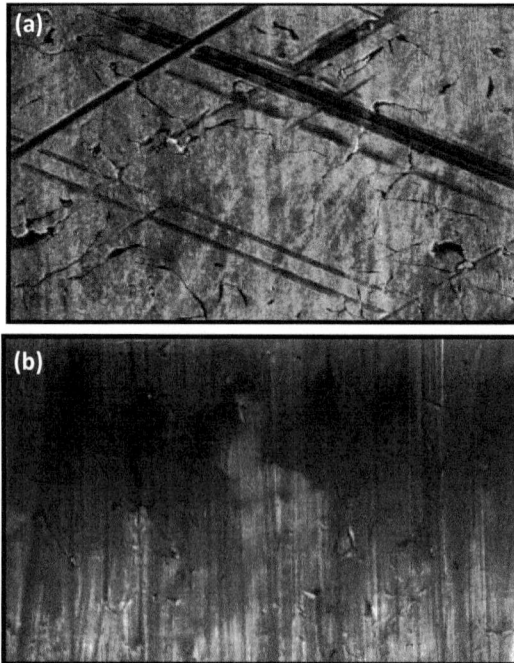

Figure 10.3 SEM photos of lab test specimens. (a) Ring, (b) cylinder.

The SEM is commonly used to generate high-resolution images of shapes of objects to show spatial variations in chemical and material compositions. Precise measurement of very small features and objects down to 50 nm in size is also accomplished using the SEM [15]. Radil [16] investigated the effects of honing on the wear of ceramic-coated piston rings and cylinder liners using the SEM technique. The baseline or control cases consisted of testing ceramic-coated rings against ceramic-coated liner specimens whose surfaces were ground and lapped smooth. The base honing process etched a distinctive crosshatch pattern on the surface of both coatings as shown in Fig. 10.3.

10.1.4 Transmission Electron Microscopy Technique

Transmission electron microscopy (TEM) is a microscopy technique in which an electronic beam is transmitted through an ultra-thin specimen, interacting with the measured specimen as it passes through reference [17]. An image is normally formed from the interaction of the transmitted electrons through the specimen; the image is magnified controlled and focused onto an imaging device, such as a fluorescent screen, on a layer of photographic film, or another detection device such as special camera.

Tribological characterization for the liner material of a diesel engine cylinder has been searched using the TEM technique to meet the requirements. Rynio et al. [18] investigated the methodology of the measurement of friction and total liner wear to monitor the formation of surface oxide layers during the experiments. The TEM technique showed some glaze layers generated at 25 and 300°C under distinct differences to reduce the wear rate by a factor of five at high temperature performance. They also searched the formation of mechanical mixing in Fig. 10.4 of oxide particles and metal matrix in a nanocomposite directly below the sliding surfaces to decrease the surface wear rate.

The imaging in the STEM represents an advanced technique to investigate the structure and composition of nanoscale materials, interfaces in the cell, and defects with atomic-scale resolution and precision. This analysis technique can help in gaining an atomic-level understanding of materials properties, which may be possible for liner material characterization [19].

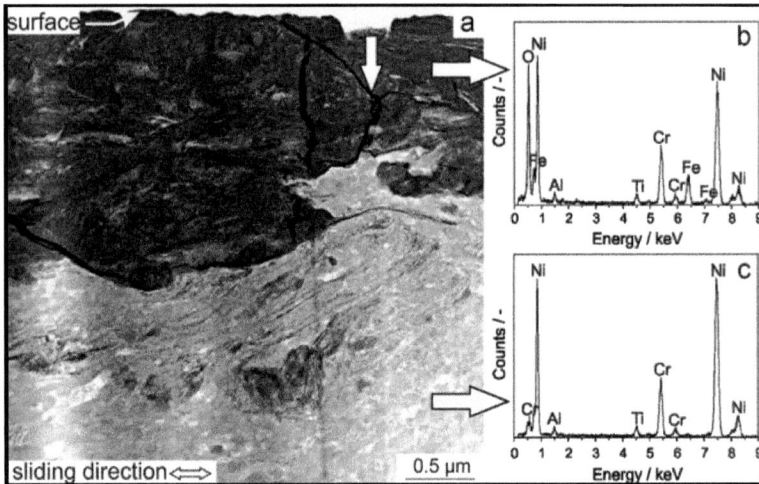

Figure 10.4 Surface microstructure analysis of cylinder material using TEM [18]. (a) Note the turbulent mixing between the glaze layer and the metal matrix. (b) EDX-spectrum from the glaze layer and (c) EDX-spectrum from the Ni-based substrate.

10.2 Tribological Behavior

Scuffing and corrosive wear as tribological problems due to sulfuric acid chemical reaction of fuel content tend to occur when diesel engines are upgraded because the conditions of operation for piston rings and the cylinder liner have become severer by upgrading [20]. Peter J. Blau [21] illustrated that the tribological behavior impacts the engine efficiency and performance of vehicles in many ways. Therefore, it is necessary to develop new design concepts of material selection, friction, wear rates and lubrication strategies for engine and drive line components. Kevin C. Radil [22] cleared that the cornerstone in the advanced heat engine development lay within overcoming the tribological problems resulting from high temperatures at the piston crown ring–cylinder liner interface; the crown ring represents the most critical part of the engine system due to its performance under high pressures and temperatures in bad boundary lubrication conditions. Consequently, the majority of the wear experienced by the piston top ring and liner occurs at this location. Therefore, lubrication tests of advanced composite materials, wear effective factors, friction analysis, and surface morphology are important tools to identify optimum combinations for enhanced engine performance and service life span.

Mehmet Caker et al. [23] studied the friction between piston ring and cylinder liner using the pin on plate reciprocating tests using Pack-boronizing method. Cast iron cylinder liner borided and measured the frictional coefficients of normal and borided cylinder liner in different reciprocating speeds. The selected specimens were cut from gray cast iron of actual cylinder liners. The chemical composition of the cylinder liner used of the experiments was given in Table 10.1. Pack-boronizing method was preferred because of the ease of treatment for boronizing of gray cylinder liner. TO conclude, the hardness of the surface has improved four times and abrasive wear resistance has increased; the roughness of the surface is decreased and the friction coefficient has also diminished.

Table 10.1 Chemical composition of the cylinder liner (%) [23]

C	Si	S	P	Mn	Ni	Cr	Mo	Cu	Fe
3.22	1.87	0.03	0.24	0.75	0.03	0.2	0.005	0.49	93.17

10.3   Coated Surface Characterization

To increase the performance quality of an internal combustion engine, the total energy must be transformed into useful energy as possible. Thermal barrier coatings of ceramics or composites are used in order to increase reliability and strength of hot metallic elements, increase yield and performance of engines and to avoid sudden breakdown. Engine coated elements with thermal barrier are piston, cylinder head, cylinder bore, and exhaust valves. Different techniques are used to coat the surfaces according to characteristics of the used materials; suitable to the intended desired use for the low cost and fuel consumption are chemical vapor decomposition (CVD), ion coating, splash coating, physical vapor decomposition (PVD), flame spray (FS), plasma spray (PS), sol-gel (SG), detonation gun (DG), electron beam evaporation coating (EBE), reactive ion coating (IP), hot izostatical press coating (HIP) from their using rates. Decrease of fuel consumption and prolonging of service life of engine elements can be satisfied by suitable thermal coating of the combustion chamber elements [24]. Funatani et al. [25] explained that the application of nickel ceramic composite (NCC) coating in two-stroke motorcycles and diesel engines has resulted in benefits in the following areas: reduced cylinder wall temperature and engine weight and increased power; lowering of fuel consumption; improved fuel economy; reduction in emissions; improved scuff and wear resistance on cylinder bores, piston and piston rings; friction reduction; reduced noise from piston slap; ability to operate in corrosive environments. The sum of these stated benefits holds much potential for contributing towards greater flexibility in materials selection for the design of light-weight fuel efficient vehicles based upon the use of aluminum engines.

Shrirao et al. [26] performed tests on a single cylinder, four-stroke direct injection diesel engine. The engine cylinder head, piston crown, and valves were coated with a 0.5 mm thickness of $3Al_2O_3 \cdot 2SiO_2$ (mullite) (Al_2O_3 = 60%, SiO_2 = 40%) upon a 150 μm thickness of NiCrAlY bond coat layer. Primary working conditions for the uncoated engine and low heat rejection (LHR) engine were kept exactly the same to initiate a comparison between the two configurations of the engine performance. They found that the standard engine heat loss to exhaust gas was about 22% higher than that of the surface-coated engine at full load. Thus, we can say, coating insulation increases the exhaust gas energy due to low surface thermal conductivity, which helps to increase the engine thermal efficiency, which increases the engine output power and decreases its specific fuel consumption.

Karuppasamy et al. [27] have used used alumina with 40% titania and nickel–chromium as thermal coating material to reduce the heat loss from the engine during operation. The coating materials TBCs also investigated by atmospheric plasma spraying technology. The same engine working conditions were maintained before and after the coating. Specific fuel consumption (SFC) and exhaust gas emissions were measured for both of conventional diesel engine compared with the same converted into Al-Ti- and Ni–Cr–coated diesel engine to determine the changes in its performance and emission characteristics of the engine. The coated diesel engine shows 16.6% lower specific fuel consumption than the standard engine. NO_x emission from the Al–Ti–coated engine cylinder is lower by 40% than the standard engine. This study shows that coating results in the increase in the brake thermal efficiency of the engine.

On the other hand, several tests were carried out to determine the performance of the thermal barrier coating (TPC) application on the compressed natural gas with the direct injection system (CNGDI) for piston crowns [28]. The experimental setup included a flame source clamped in front of the piston crown sample to expose the piston crown surface to direct heat, and the flame power was kept fixed during the test to ensure better temperature control, as shown in Fig. 10.5.

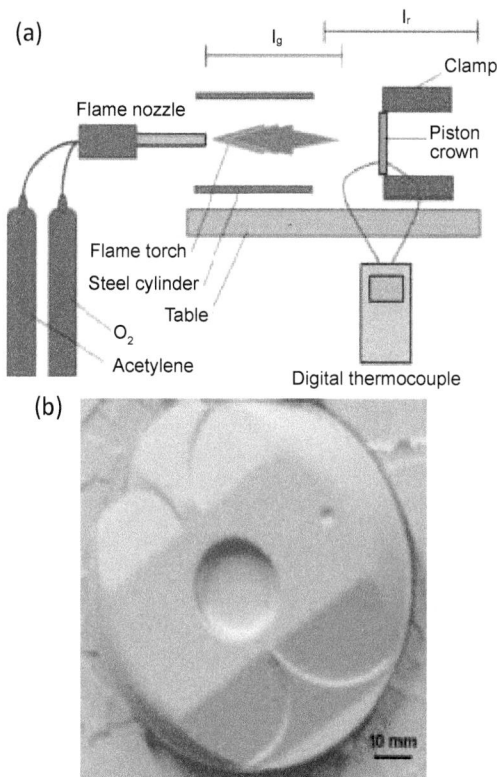

Figure 10.5 Experimental setup [28]: (a) Burner rig test in horizontal view. (b) Actual plasma-sprayed YPSZ-coated piston crown.

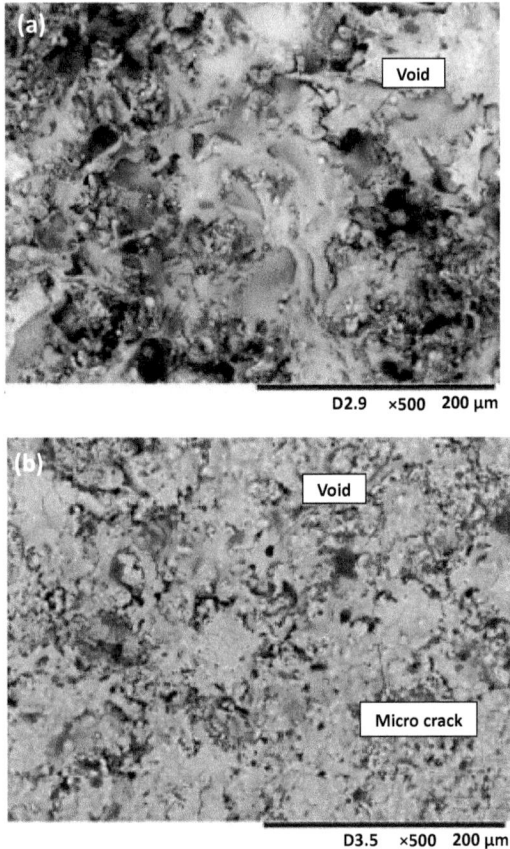

Figure 10.6 Microstructure characterization of coated top surface by TEM: (a) Plasma-sprayed NiCrAl and (b) plasma-sprayed YPSZ [28].

Figure 10.5b shows the plasma-sprayed YPSZ-coated piston crown that has surface roughness of about 9.2 μm and microhardness of 762.3 HV. The YPSZ/NiCrAl-coated samples were cut into small pieces for necessary quantities. After that, the samples were washed with acetone and were dried before the cross sections of the samples were polished. Then, the pieces of polished sample were hardened and mounted in the mixture of epoxy resin and epoxide hardener for metallographic examination. The microstructure images were taken for the surfaces of bond coat NiCrAl, topcoat YPSZ, and its cross-sectional view. The samples of plasma-sprayed YPSZ $(7.5Y_2O_3ZrO_2)$/NiCrAl-coated piston crowns were observed for the surface structure using an SEM technique. The microphotograph of the fracture surface on the cross-sectional view of the piston crown samples represents the microstructure of the top surface of plasma-sprayed NiCrAl bond coating and ceramic-based YPSZ coating, respectively (see Fig. 10.6). The particles of both materials were deformed on impact during the plasma-spraying process and melted on the piston crown surface. The surface structure of the NiCrAl bond coating had a bigger dense splat like pattern and a few of big voids which showed low porosity. Compared to the ceramic-based YPSZ coating, the structure of the surface showed fine particles with a lot of small voids, which means high porosity and numbers of micro cracks on the surface. The high porosity characteristic of TBC might be the reason for low thermal conductivity that reduced the heat transfer by conduction between the combustion chamber to the engine piston.

From the other viewpoint, the anti-wear additive zinc dialky dialkyl dithiophosphate (ZDDP) was essentially discovered accidently in the 1940s [29]. With new analysis techniques, researchers discovered that ZDDP breaks down and turns into a "tribofilm," a thin, solid layer that adheres to the surfaces in contact and further protects them from wear (see Fig. 10.7) [29].

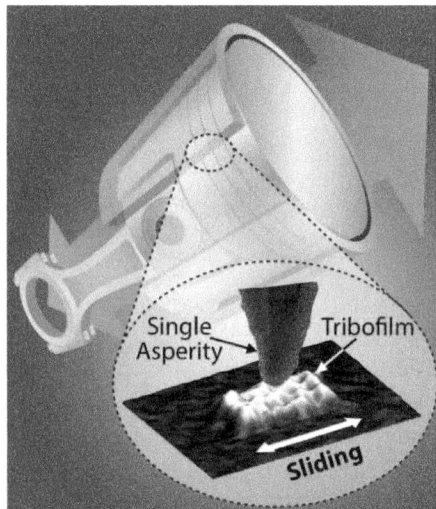

Figure 10.7 Surface microstructure of ring material using AFM technique [29].

10.4   New Applied Technology in Engine Coating Surfaces

Technology for coating the inside of the engine has existed for a long time. New technologies in surface coating, especially in rubbing rotating elements of engine materials have been applied commercially. The main purpose of this technology is to reduce friction, noise, engine weight, and fuel consumption. Surface coating using diamond-hard technology has being investigated for engine piston ring as shown in Fig. 10.8.

A company has developed a range of piston coating technologies designed to support heavy-duty diesel engine manufacturers in reducing fuel consumption and meeting the emission requirements (Fig. 10.9a) [31]. The advanced process uses a controlled re-melting of the bowl rim and/or the bowl base to refine the grain size of the silicon to just one-tenth of its original size, significantly extending the fatigue life of a cast piston. Figure 10.9b shows the result of a unique process using high-precision electro-erosion machining (HPEEM), which removes an exact amount of material, thus precisely shaping the piston pin bore, enabling improved load capacity and reliability.

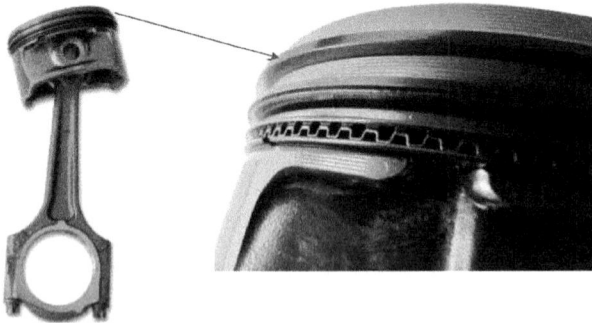

Figure 10.8   Coated piston ring using Diamond [30].

Another company investigated engine cylinder liners, through the mirror bore coating technology by spraying molten iron into the surface of the cylinder bore, forming an iron coating layer on the walls inside. They were able to utilize the technology for some of its vehicles. Through the unique technology that pretreats the aluminum surface by spraying molten iron, R&D could achieve a large cost reduction.

Figure 10.9 Coated engine pistons [31]. (a) Aluminum piston coated by silicon (gasoline engine). (b) Steel piston coated by nanodiamonds in a chrome layer for piston skirt (diesel engine).

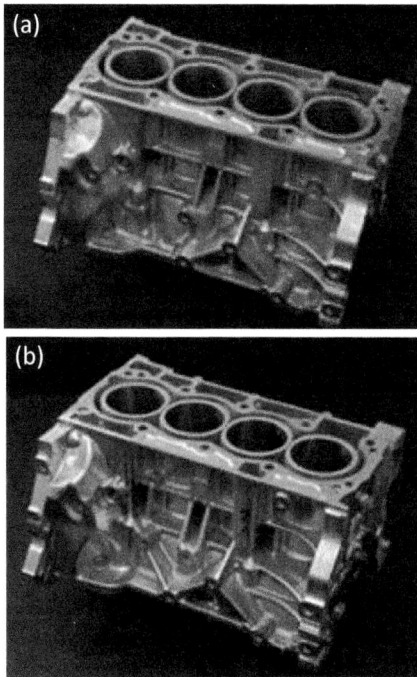

Figure 10.10 Four-cylinder engine block [32]. (a) A cast iron liner in conventional engine block. (b) A new type of with mirror bore coating in new engine block.

10.5   Machining Characteristics of Engine Cylinder Surface

The central point of the machining work takes place on the automotive engines happens at the computerized machining center. Mezghani et al. [33] discussed various forms of the wear modeling that caused running-in problems in a honed cylinder surface and its implications on ring-pack friction response. Plateau honing experiments under different surface finishing conditions were carried out on a vertical honing machine with an expansible tool (NAGEL No. 28–8470), as shown in Fig. 10.11. The plateau honing characterizes the surface wear profile during the running-in period of cast iron engine cylinders using advanced characterization techniques. The smooth surfaces produced less friction response despite the increase in the ratio between plateau and valley height.

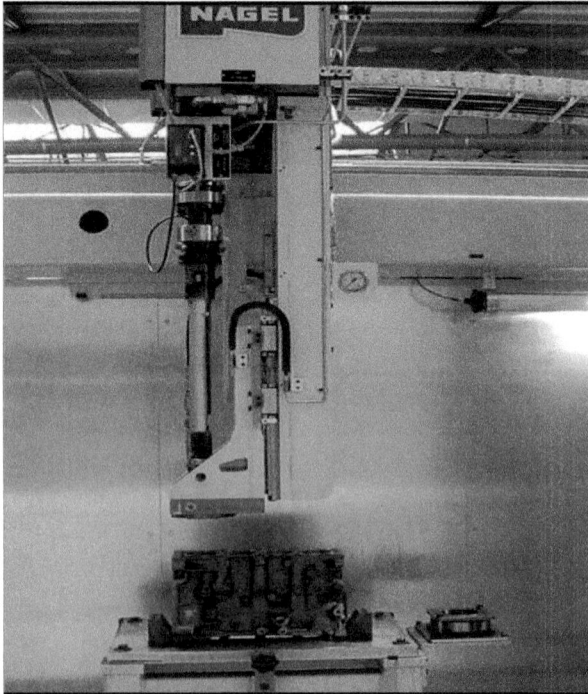

Figure 10.11 A honing machine with expansible tool for cylinder block [33].

Another CNC machine type (RMC V30 4-axis) is capable of handling every aspect of automotive engine blocks (see Fig. 10.12). The CNC block machining center is able to correct the factory machine work and ensure that the bores and machined surfaces are perfect. The accuracy of the machined surfaces is exactly according to the factory blueprint locations ±0.0001. Factory Ford blocks often measure ±0.002 from the blueprint.

Figure 10.12 Honing surface for cylinder using RMC V30 4-axis CNC [34].

Veeresh Kumar et al. [35] attempted to consolidate some of the aspects of mechanical, wear behavior, and prediction of the mechanical and tribological properties of aluminum base MMC. Plotted aluminum MMCs are sought over other over other conventional materials in the field of aerospace, automotive, and marine applications owing to their excellent improved properties. These aluminum MMC composites initially replaced cast iron and bronze alloys, but owing to their poor wear and seizure resistance. It was discovered that as the reinforcement contents increased in the matrix material, the hardness of the composites also increased.

The surface roughness of the measured cylinder honing profile is presented in a 2D image using probe scanning stylus profilometry technique (see Fig. 10.13). The surface characteristics of the cylinder honing profile are measured and presented in a 3D image using optical white light technique (see Fig. 10.14). The 3D steep slopes can show additional height, and light can be reflected on the surface and no data can be obtained.

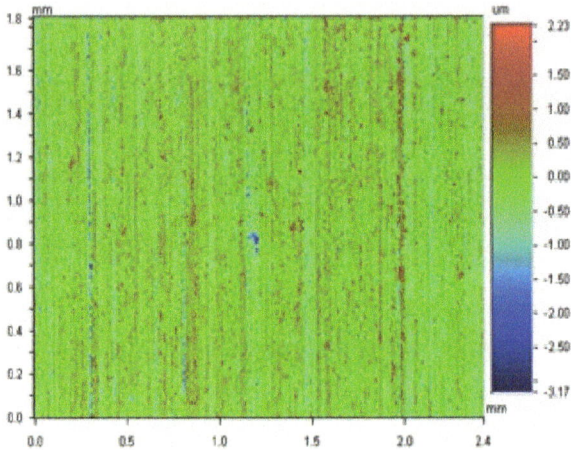

Figure 10.13 Honing surface profile for cylinder using stylus profilometry [36].

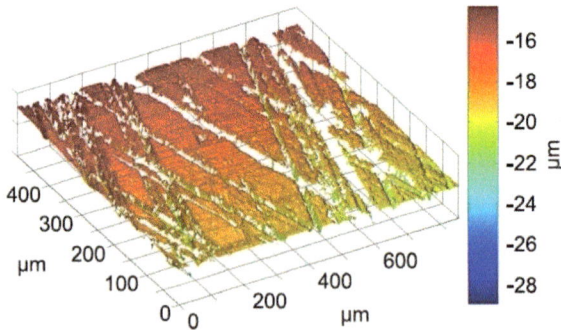

Figure 10.14 Honing surface for cylinder using white light interferometry [36].

10.6 Conclusion

From the analyses in this chapter, the following conclusion can be drawn:

- The quality of engine performance and its service life span are functions of the cylinder, piston, and piston rings material, which define the friction losses and surface-specific wear rate.
- Each material's surface quality depends on the size of the used material cell form and dimensions, which control the surface roughness and cell rigidity against friction and wear.
- Advanced metrology techniques at micro- and nanometer scales play an important role in the characterization of developed surfaces for automotive engine materials in both honing and composition conditions.
- Surface finishing can be limited according to the surface material characteristics. Consequently, each material or surface coating type for engine elements has its limited surface finishing, which controls the tribological characteristics.
- Surface finishing quality of engine rubbing rotating elements requires suitable lubricant characteristics to minimize friction losses. Therefore, accurate surface topography classifications ensure a good choice of the requirements of engine lubricant characteristics.
- In spite of today's research, a compromise is still required between the advantages of metallic and nonmetallic materials in the automotive engine materials. This will be easily achieved using non-metallic materials in the coating of metallic material to develop the engine performance. Consequently, experimental research is required to compare the available techniques of engine surface materials, coating, and surface finishing quality.

References

1. European Aluminium Association, *Applications-Power Train-Cylinder Linings.* The Aluminium Automotive Manual, 2011.

2. A. Fakaruddin, A. Hafiz, and K. Karmegam, Materials Selection for Wet Cylinder Liner, *AOSR Journal of Engineering (IOSRJEN)*, vol. 2, no. 9, pp. 23–32, September 2012.

3. J. R. D. Davis, Friction and Wear of Internal Combustion Engine Parts, *ASM Handbook, Friction & Lubrication and Wear Technology, ASM International*, vol. 18, pp. 553–562, 1992.

4. N. Hieu, Manufacturing Processes and Engineering Materials Used in Automotive Engine Blocks, School of Engineering Grand Valley State University, April 8, 2005.

5. J. Lenny Jr, Replacing the Cast Iron Liners for Aluminum Engine Cylinder Blocks: A Comparative Assessment of Potential Candidates, *Master of Engineering in Mechanical Engineering*, April 2011.

6. F. Manabu, K. Shinji, T. Toshihiro, and H. Shigeru, Development of Metal Matrix Composite for Cylinder Block. FISITA World Automotive Congress, Seoul, 2000.

7. L. G. Wilson, V. D. Pedro, and K. Walmor, Compacted Graphite Iron for Diesel Engine Cylinder Blocks. *Brazilian MRS Meeting, Congrès Le diesel: aujourd'hui et demain.* 1–11, March 2004.

8. R. Parkinson, Properties and Applications of Electroless Nickel. Nickel Development Institute. Online website at: www.google.com.

9. I. E. Ayoub, Study of Electroless Ni-P Plating on Stainless Steel, *The Online Journal on Mathematics and Statistics (OJMS)*, vol. 1, no. 1, 2014.

10. S. H. R. Ali, Advanced Measuring Techniques in Dimensional and Surface Metrology. *X*th *International Scientific Conference Coordinate Measuring Technique*, Bielsko-Biała, April, 23–25 2012.

11. S. H. R. Ali, Performance Investigation of CMM Measurement Quality Using Flick Standard, *Journal of Quality and Reliability Engineering*, vol. 2014, ID 960649; pp. 1–11, 2014.

12. S. H. R. Ali, CMM in Automotive Metrology. *X*th *International Scientific Conference coordinates Measuring Technique*, Bielsko-Biała, April, 23–25, 2012.

13. S. H. R. Ali. Advanced Nanomeasuring Techniques for Surface Characterization, *International Scholarly Research Network*, ISRN optics Article. 2012, ID 859353, 2012.

14. J. A. Last, P. Russell, P. F. Nealey, and C. J. Murphy, The Applications of Atomic Force Microscopy to Vision Science, *Investigative Ophthalmology Visual Science, ARVO Journal*, vol. 10, p. 5470, 2010.

15. S. Susan, Scanning Electron Microscopy (SEM), Geochemical Instrumentation and Analysis, University of Wyoming, Carlton College, 2012.

16. R. Kevin, The Influence of Honing on the Wear of Ceramic Coated Piston Rings and Cylinder Liners, U.S. Army Research Lab., Glenn Research Center, Cleveland, Ohio, 2000.

17. Z. Pavel, Transmission Electron Microscope, Advanced Techniques in Geophysics and Materials Science, Lecture 14, University of Hawaii, Honolulu, USA.

18. C. Rynio, H. Hattendorf, J. Klöwer, and G. Eggeler, On the Physical Nature of Tribolayers and Wear Debris after Sliding Wear in a Superalloy/Steel Tribosystem at 25 and 300 C, *Wear*, vol. 317, no. 1–2 pp. 26–38, 2014.

19. N. D. Browning, J. P. Buban, M. Chi, B. Gipson, M. Herrera, D. J. Masiel, S. Mehraeen, D. G. Morgan, N. L. Okamoto, Q. M. Ramasse, B. W. Reed, and H. Stahlberg, The Application of Scanning Transmission Electron Microscopy (STEM) to the Study of Nanoscale Systems, *Modeling Nanoscale Imaging in Electron Microscopy, Nanostructure Science and Technology*, vol. 9, p. 182, 2012.

20. S. Saburi, Y. Saitoh, and T. Yamada, Tribology between Piston Rings and Cylinder Liners of Marine Diesel Engines, *IHI Engineering Review*, vol. 38, no. 1, 2005.

21. P. J. Blau, A Review of Sub-Scale Test Methods to Evaluate the Friction and Wear of Ring and Liner Materials for Spark and Compression Ignition Engines, U.S. Department of Energy, 2001.

22. K. C. Radil, Test Method to Evaluate Cylinder Liner-Piston Ring Coatings for Advanced Heat Engines, *Prepared for the 51st Annual Meetings sponsored by the Society of Tribologists and Lubrication Engineers Cincinnati*, Ohio. May 19–23, 1996.

23. M. Caker and I. hakki, Frictional Behavior between Piston Ring and Cylinder Liner in Engine Condition with Application of Reciprocating Test, *International Journal of Materials Engineering and Technology*, vol. 11, no. 1, pp. 57–71, 2014.

24. G. Hüseyin and G. Hasan, An Evaluation of the Effects of Coating with Thermal Barrier on Engine Performance in Diesel Engine, *Technology*, vol. 13, no. 1, pp. 49–57, 2010.

25. K. Funatani, K. Kurosawa, P. Fabiyi, and M. Puz, Improved Engine Performance via Use of Nickel Ceramic Composite Coatings (NCC Coat), *SAE Technical Paper No. 940852*, 1994.

26. P. N. Shrirao and A. N. Pawar, Experimental Investigation on Performance of Single Cylinder Diesel Engine with Mullite as Thermal Barrier Coating, *Journal of Mechanical Engineering and Technology*, ISSN: 2180-1053. vol. 3, no. 1, January-June 2011.

27. K. Karuppasamy, M. P. Mageshkumar, T. N. Manikandan, J. Naga Arjun, T. Senthilkumar, B. Kumaragurubaran, and M. Chandrasekar, The Effect of Thermal Barrier Coatings on Diesel Engine

Performance, *ARPN Journal of Science and Technology*, vol. 3, no. 4, April 2013.

28. H. A. Jalaludina, S. Abdullahb, M. J. Ghazalib, B. Abdullahc, and N. R. Abdullahc, Experimental Study of Ceramic Coated Piston Crown for Compressed Natural Gas Direct Injection Engines, *Procedia Engineering*, vol. 68, pp. 505–511, 2013.

29. E. L. Penn and E. Mobil, Researchers Address Long-standing Mysteries Behind Anti-wear Motor Oil Additive, March 2015.

30. Diamond Rings Shine on Chevrolet Sonic, Cruze GM, http://www.gminsidenews.com/forums/f13/diamond-rings-shine-chevrolet-sonic-cruze-112136/, July 10, 2012.

31. N. Romain, New Developments In Heavy-Duty Diesel Pistons, 09-4-2014. http://www.car-engineer.com/new-developments-heavy-duty-diesel-pistons/.

32. Future Technology, Nissan Motor Coroporation. http://www.nissan-global.com/EN/TECHNOLOGY/OVERVIEW/mirror_bore_coating.html.

33. S. Mezghani, I. Demirci, M. Yousfi, and M. El Mansori, Running-in Wear Modeling of Honed Surface for Combustion Engine Cylinder Liners, *Wear*, vol. 302, pp. 1360–1369, 2013.

34. TRM V30 CNC, Livenois Co. http://www.livernoismotorsports.com.

35. G. B. Veeresh Kumar, C. S. P. Rao, and N. Selvaraj, Mechanical and Tribological Behavior of Particulate Reinforced Aluminum Metal Matrix Composites: A Review, *Journal of Minerals & Materials Characterization & Engineering*, vol. 10, no. 1, pp. 59–91, 2011.

36. Cylinder liner honing angle texture and procedure, Dev. 19, 2014. http://hfoplant.blogspot.com/2014_12_01_archive.html.

PART 6
CONCLUSIONS AND FEEDBACK FOR FUTURE

Chapter 11

Conclusions, Recommendations, and Future Work

In recent decades, metrology has achieved a great revolution of high quality in the automotive industry. The high quality of the final automotive engine product or maintenance requires accurate and precise metrology techniques. In fact, metrology played an important role of accurate and precise technology in improving the performance of automotive engines. From the scientific results presented in this book, the author confirmed that the engine metrology became the cornerstone in modern automotive industries.

Nowadays, the measurement quality with the level of automation has increased the degree of measurement accuracy in manufacturing parts, especially in an automotive engine. The quality philosophy of automotive engine parts designates the processes of measuring their dimensional and geometrical features through important processes for either inspection or diagnoses. The purpose of these measurement techniques is to assess and ensure the compliance of the engine parts with the intended design specifications.

Particularly in the modern automotive engines, dimensional and geometrical form inspection constitutes a dominant portion of the total assessment work. The neighboring engine parts with mating surfaces in relative motion, either rotating or reciprocating,

Automotive Engine Metrology
Salah H. R. Ali
Copyright © 2017 Pan Stanford Publishing Pte. Ltd.
ISBN 978-981-4669-52-8 (Hardcover), 978-1-315-36484-1 (eBook)
www.panstanford.com

are considered the most vital parts as they are practically responsible for the friction losses, energy dissipation due to friction resistance, as well as the associated wear occurrence. Dimensional and geometrical distortion results due to wear on those working surfaces and subsequent continuous energy and pressure languish in the engine to the limit of urgent demand for replace or overhaul. Actually, the most important goal of dimensional measurement is to introduce the best efficiency and ensures the continuity of research and development (R&D) to improve automotive engine technology. Therefore, automotive engine manufacturers should own and primarily rely on coordinate measuring machines (CMMs) because of their extraordinary ability to work with easy and complicated parts.

Important challenges called "measurement quality" that face automotive manufacturer to define data quality key performance indicators actress in completeness, validity, integrity, timeliness and measuring consistency may be relatively easy to measure, but accuracy and precision together with uncertainty are a whole new story. As it is well known, Fig. 11.1 illustrates the relation between the difficulty of measurement quality and the business impact.

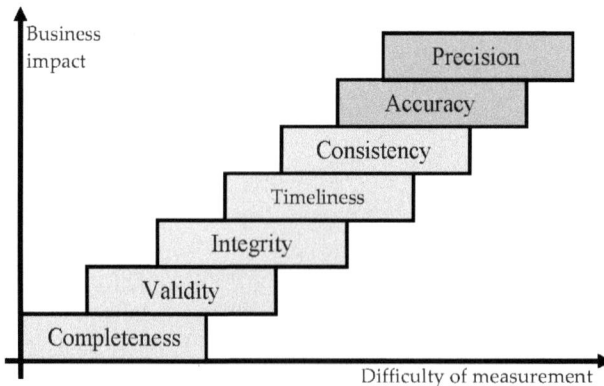

Figure 11.1   Environment of measurement quality.

11.1   Conclusions

This book summarizes and discusses the roles, motivations, and update of instrumentation technology and measurement strategies, including errors and uncertainty issues in automotive engines. The following conclusions can be drawn:

- Accurate measurement of the dimensional and geometrical form for automotive engine parts is a matter of great importance in the modern auto manufacturing technology.

- Accuracy and precision in measurement for fits and tolerance results in great harmony, especially in modern automotive engine industries.

- If you have new coordinate techniques, you can implement a comprehensive inspection program to exploit the advanced precise dimensional and geometrical measurements in automotive engine repair and overhaul to be successful in predicting the engine performance. So, crucial engine parts that dominantly affect the engine performance, such as cylinder liners, pistons and rings, and valves and seats, can be inspected successfully.

- I wish to see a permanent development in the community culture level and technical awareness among professionals in industry and students in specialized educational institutes and universities (production engineering, precision engineering, and automotive engineering departments) for the importance of dimensional and geometrical measurements in automobile engines.

- A marked improvement is expected in the future with respect to the explanation and analysis of geometrical features, tolerances, and orientations, including the results of dimensional measurements for the benefit of the auto engine industry and maintenance.

- Eventually, as a call for creativity, I invite all scientific researchers, engineers, and experts for scientific cooperation for the invention of a new optic-electro-mechanical instrument offering higher accuracy in dimensional measurements.

11.2 Recommendations

Those who want to progress in the automotive engine industry must accelerate direction to rely on excellent coordinate measuring machines besides the qualification of engineers. Moreover, it is important to habilitate of their workers and users. It is also very important establish a scientific section for research and development (R&D). Eventually, government support for ministries of industry and trade is a prerequisite for success.

11.3   Future Work

The author, with collaboration with sincere researchers and scientists, hopes to harness light and optics to develop a new optic-electro-mechanical instrument for a more accurate metrology technique with a larger volumetric measurement range. Moreover, a new technique can be developed by the combination of the AFM technology and the CMM technology. This will lead to a substantial shift in the automotive engine industrial technology.

Index

accuracy 6–7, 14, 31, 38, 42–43, 98, 100, 107, 127–128, 130, 153, 184–185, 199–200, 261, 270–271

AFM, *see* atomic force microscope

algorithm techniques 194–195, 197

algorithms 67, 130, 149, 158, 163, 170–174, 177, 184, 194–195, 249

  surface detection 68

Al–Ti-coated engine cylinder 255

aluminum alloy 246

artifacts 15–16, 90–92, 109–110, 117, 188, 199

atomic force microscope (AFM) 49–51, 60, 63, 68, 249

automotive engine blocks 261

automotive engine cylinder liner 58

automotive engine parts 269

automotive engine repair 225, 271

automotive engines 4–5, 9–10, 13, 17–18, 21, 203, 245, 248–249, 260, 269, 271–272

automotive industry 3, 8, 14, 28, 58, 83, 125, 247, 269

automotive metrology 9

average roundness accuracy 199

BDC, *see* bottom dead center

bottom dead center (BDC) 220, 223, 230–231, 238, 240–241

Burner rig test 255

calibration 13–14, 16, 54, 103, 132, 154, 211

CCD, *see* charge-coupled device

CCD camera 49, 207–208, 210

ceramic-based YPSZ coating 257

CGI, *see* compacted graphite iron

charge-coupled device (CCD) 31–32, 39, 207

chemical analysis 234–235

chemical energy 17–18

chemical vapor decomposition (CVD) 253

circle fitting techniques 191

CMM, *see* coordinate measuring machine

CMM accuracy 29, 126, 130–131

CMM error signals 119

CMM errors 29, 119–120

CMM fitting algorithms 129, 176

CMM measurements 6, 9, 29, 86, 103, 108–109, 117, 121, 155, 158, 208, 232

CMM metrology technique 81, 247

CMM performance 5, 7, 177, 248
CMM performance accuracy
    154–155
CMM probe ball tip error 87
CMM probe styli repeatability
    tests 93
CMM probe stylus 87, 93
CMM probes 83–85, 131, 156
CMM scanning probing error
    131
CMM setup measurements
    strategy 211
CMM software 7, 109, 129–130,
    154
CMM structure errors 120
CMM stylus systems 91, 108,
    118, 131
CMM test element specification
    131
CNTs 61
coated surface characterization
    253, 255, 257
compacted graphite iron (CGI)
    247
compression pressure 212, 216,
    219
  engine cylinder peak 224
compression pressure tests 224
computed tomography (CT) 4,
    28, 52
coordinate measuring machine
    (CMM) 4–9, 28–33, 58,
    66, 83–86, 96, 107–108,
    127, 130–132, 153–156,
    175–178, 183, 207–211,
    233–234, 270–271
CT, *see* computed tomography
CVD, *see* chemical vapor
    decomposition
cylinder block 208, 212, 221,
    246, 260

cylinder block attributes 212
cylinder block measurements
    212
cylinder blocks 246
cylinder bores 9, 218, 222, 249,
    254
cylinder configuration 234
cylinder head 18, 212, 223–224,
    253–254
cylinder liner 60
cylinder liners 58–60, 208–209,
    215, 219–220, 222, 225,
    241, 250–253, 271
cylinder material 234, 252
  surface microstructure analysis
    of 252
cylinder out-of-roundness 233
cylinder roundness
    measurements 127
cylinder straightness 216, 239
cylinder surface
  honed 260
  smooth 58
cylinder wall 230–231, 233,
    245, 247
cylinders
  cast iron 260
  worn-out 240

design 7, 17, 19, 33, 35, 85, 119,
    154, 175
detonation gun 253
deviations
  geometric 241–242
  geometrical form 220, 222
  roundness form 157, 183
  surface-profile 186
diameter error 158, 160–162,
    164, 167, 171

average variation of 159, 164, 167

diameter error limits 159, 161, 164, 167

diameter error variations 167

diameter errors variation of LSQ fitting algorithm 159

diameter errors variation of MC fitting algorithm 165

diameter errors variation of MI fitting algorithm 167

diameter errors variations 162

diesel engine cylinders 251

  air-cooled 234

diesel engines 9, 18–22, 232, 247, 252, 254, 259

digital holography 48–49

digital holography technique 48

dimensional measurements 9, 53, 213, 218, 225, 229, 232, 241, 270–271

dimensional metrology 7–9

dynamic errors 87, 90, 98

dynamic resonance 113, 116, 121

Egyptian royal cubit 10

engine liners 218–219

engine piston 60, 257–258

engine wear 9

engine weight 254, 258

enhanced vertical scanning interferometry (EVSI) 40

error sensitivity 119

errors

  geometric 193–194

  high amplitude 116

  probe detection 85

  probe tip 66

propping 131, 155

relative deviation 155

roundness deviation 158

EVSI, *see* enhanced vertical scanning interferometry

exhaust valve 210, 213–214, 253

fitting algorithm application 128

fitting algorithm response 148, 170, 172

fitting algorithm techniques 193

fitting algorithm types 132

fitting algorithms 7–8, 127, 130, 133–146, 148–149, 154, 158, 161–164, 169–171, 173, 177, 187, 190, 196, 198–199

  evaluation 147, 158

  geometric 153–155, 176

  LS 192

  LSQ 133, 159–160, 175

  MC 164–165

  MI 167–168

Flick artifact transverse circle 161, 164, 167

Flick diameter 170–171

Flick diameter error 171

Flick roundness 172–173

Flick roundness error 174

Flick transverse circle 160, 162, 165, 168

Fourier analysis 6, 111, 121

friction 96, 230, 240, 242, 249, 251–253, 258, 263

friction losses 205, 230, 246, 263, 270

fuel consumption 253–254, 258

fuel economy 21

gasoline 18
Gaussian filter 190–195, 198, 200
Gaussian filtering technique 57
Gaussian fitting technique 199
geometrical product specification (GPS) 5, 27
GPS, *see* geometrical product specification
gravity 19, 231

high-precision electro-erosion machining (HPEEM) 258
HPEEM, *see* high-precision electro-erosion machining

ICEs, *see* internal combustion engines
interferometers 42–43, 67
internal combustion engine cylinders 246
internal combustion engines (ICEs) 10, 18, 21–22, 245, 253

laser beam 37, 42, 50–51
laser scanning microscope (LSM) 61, 63
least square 130, 132, 158–161, 187, 189, 194–196, 233
LSM, *see* laser scanning microscope
LSQ algorithm 146, 148, 160, 172
lubrication tests 253

machine root error 114, 116, 120
maximum inscribed element (MIE) 133–141, 143, 145–148
MC, *see* minimum circumscribed
measurement errors 7, 87–88, 112, 116, 120, 129, 153, 156, 188–189, 248
  dynamic 85
  permissible 91, 131
measurement standards 12–13
metal matrix composite (MMC) 246
metrology
  applied 13
  industrial 13
  scientific 13–14
metrology engineers 7, 186
metrology techniques
  accurate 272
  advanced 3, 25, 27, 263
  advanced soft 4–5, 107, 247
  non-contact 68
  precise 14–15, 269
micro coordinate metrology 153
microscopy, optical 4, 28, 41, 44
MIE, *see* maximum inscribed element
MIE fitting 138, 142, 144
MIE fitting error 142, 144
minimum circumscribed (MC) 133, 158, 164–166, 170–176, 188–189, 194–196
MMC, *see* metal matrix composite
multiple wavelength interferometry (MWI) 40
MWI, *see* multiple wavelength interferometry

NA, *see* numerical aperture
nanocomposites 61–62, 251
NCC, *see* nickel ceramic
    composite
nickel ceramic composite (NCC)
    254
non-optical measurement
    techniques 28, 49, 51, 53
numerical aperture (NA) 43

on-machine roundness
    measurement 36
optical microscopes 5, 55, 207,
    210

peak roundness accuracy 198
peak roundness error 193–194,
    196
phase-shifting interferometry
    (PSI) 39–40, 42, 45
photodiode 37
physical vapor decomposition
    (PVD) 253
piston crown 255, 257
  plasma-sprayed YPSZ-coated
    255–256
piston rings 230, 233, 246, 249,
    252, 254
piston stroke 234, 236
piston velocity 230–231
plasma spray (PS) 61, 253
polyvinylchloride (PVC) 61
power strokes 230, 238, 240
precision 7–8, 14, 107, 125, 184,
    212, 251, 270–271
probing error 131, 155, 211
  permissible 91, 131, 156, 177
production engineering 271

PS, *see* plasma spray
PSI, *see* phase-shifting
    interferometry
PVC, *see* polyvinylchloride
PVD, *see* physical vapor
    decomposition

resolution 31, 38, 40, 43–44, 47,
    49, 54–55, 111, 175
rotation 86, 233
rotating elements 8, 245, 258,
    263
rotating gauge 33, 34
rotating parts 183, 185
rotating table 33, 183
rotational factor 128
root errors 89–90, 108, 121
  dynamic 6, 86
  systematic 6
roughness 38–39, 45, 54, 57, 63,
    65, 86, 128, 209, 253
roughness measurements 38–39,
    63
roundness 8, 33, 36, 127–128,
    133, 140–141, 146, 149,
    169–170, 172–173, 183,
    195–196, 212–213, 216,
    234–238
roundness accuracy 189, 197,
    199
roundness accuracy development
    193
roundness accuracy
    measurement 190
roundness deviations 8, 58, 127,
    155, 220–221
roundness error 101, 132–133,
    135–144, 147–149,
    160–166, 168–169, 173,
    188–189, 191, 194, 196, 248

roundness error limits 134–136, 140
roundness error variation 135–136, 163
roundness form metrology 185
roundness measurement accuracy 8, 148, 184, 191, 199–200
roundness measurement error 7, 127, 142, 145, 148, 161, 163, 166, 169, 200
roundness measurements 7–8, 34, 36–37, 144, 149, 158, 163, 172–173, 176–177, 183–185, 187–188, 190–193, 195–196, 200, 209
roundness metrology 35, 154, 184, 191, 197, 199
roundness nanometrology 8

scanning confocal microscope 44–45
scanning electron microscopes (SEM) 45, 47, 250–251
scanning electron microscopy 46
scanning probe microscope 45, 49
  advanced 249
scanning probe microscopy (SPM) 4, 28, 45, 49–50, 249
scanning probing error
  permissible 132, 149, 156
  permissible tangential 91, 211
scanning transmission electron microscopy (STEM) 251
scanning tunneling microscope (STM) 49–50
SD, *see* standard deviation

seizure resistance 59, 261
SEM, *see* scanning electron microscopes
sensitivity coefficients 148–149, 172, 174
SFC, *see* specific fuel consumption
SG, *see* sol-gel
silicon 247, 258–259
silicon wafer 63
smooth surfaces 45, 260
sol-gel (SG) 253
specific fuel consumption (SFC) 254–255
spindle error 36, 120, 189
spiral sampling 39–40
SPM, *see* scanning probe microscopy
standard artifact surface topography 102
standard deviation (SD) 93, 95–97, 131, 145–146, 155–156, 169–171, 173, 217–218, 231, 235
STEM, *see* scanning transmission electron microscopy
STM, *see* scanning tunneling microscope
straightness 128, 208, 216, 218, 235–237, 240
straightness deviation 220–221
straightness measurements 209, 234–236
stylus 6, 29, 56, 83, 85–87, 89, 95–96, 98, 108–111, 114, 116–121, 185, 189–191
  scanning speeds of 109, 111, 117
stylus ball 86–87, 89
stylus design 93, 109, 117
stylus detection 117–119

stylus errors 6, 107–108, 111,
    118
stylus root errors 111, 117, 121
stylus specifications 110–111,
    190
stylus tip 6, 83, 86, 95–96, 99,
    184
stylus tip diameter 112, 121
stylus tip radius 6, 94–95, 97, 99
surface coating 258
surface deformation 87
surface finishing conditions 260
surface finishing quality 263
surface inspection techniques
    207
surface metrologists 57
surface metrology 5, 8, 10, 27,
    44, 51, 57, 61, 68
surface metrology techniques
    28, 61
surface microstructure of ring
    material 257
surface morphology 65, 253
surface nanometrology 27–28
surface scanning 50, 117
surface texture 27, 57–58
surface topography 27, 39, 45,
    56–58, 209, 231, 245
surface waviness 98, 103
surface wear rate 251
systematic errors, quasi-static
    85

TDC, *see* top dead center
TEM, *see* transmission electron
    microscopy
thermal barrier coatings 255
three-way catalyst (TWC) 22
top dead center (TDC) 220, 223,
    230–231, 234, 238, 240

transmission electron microscopy
    (TEM) 251, 256
TWC, *see* three-way catalyst
two-trigger-stylus, dynamic
    amplitude response of 112,
    116

uncertainty 6, 30, 37, 66–67,
    96–97, 126, 154–155, 171,
    173, 175–177, 207,
    216–219, 235, 270
  constant machine 131, 156
undulations 6, 121, 156, 187,
    194–195, 199, 248
  cylindrical surface form 6
unforeseeable errors 6, 87, 89

V-type engine 19–20
valve seats 9, 207–208, 213–214,
    249
valves 9, 206–208, 210,
    213–214, 222, 225, 249,
    254, 271
VAST scanning probe head 91
VAST scanning stylus 111
vertical scanning interferometry
    (VSI) 39–40
VSI, *see* vertical scanning
    interferometry

waviness 33, 39, 57, 91, 128, 247
waviness profile 98, 100, 103,
    110
wear 9, 60, 153, 205, 229–231,
    233, 240–242, 251–253,
    257, 261, 263, 270

wear rates   251–252, 263
white-light interferometer
   (WLI)   28, 43, 45, 60
white light system   42–43

WLI, *see* white-light
   interferometer
WLI microscopy   42